Coral Triangle Cameos

Biodiversity and the Small Majority

Hard coral, *Acropora* sp., colour variations

Coral Triangle Cameos

Biodiversity and the Small Majority

Alan J Powderham
Photography, text, graphic design

Sancia ET van der Meij
Scientific consultant

DIVED UP

ISBN 978-1-909455-57-3 (Hardback)

Cataloguing-In-Publication Data. A catalogue record
be obtained from the British Library.

Contents

About the author

Alan J Powderham has been a scuba diver and underwater photographer for over forty years. By profession he is an engineering consultant specialising in sustainable solutions and environmental engineering, promoting green and low carbon technology. His previous books are: *Venezuela Submarina*, on the marine life of the southern Caribbean (1983, Oscar Todtmann); and *At the Heart of the Coral Triangle* (with Sancia ET van der Meij) (2021, CRC Press).

Scientific consultant

Dr Sancia ET van der Meij is a leading marine biologist with extensive field experience in the Coral Triangle. Passionate about unravelling the secrets of coral reefs, she heads her own lab at the University of Groningen dedicating most of her time to the small majority. The tiny invertebrates living symbiotically with a host organism are at the core of her research, and she has a soft spot for minute coral-dwelling gall crabs.

Regarding organism identification

While a great deal of time and care has been devoted to making accurate classifications, identifying species from photographs alone will always involve a degree of uncertainty. As such, species names in this book should be treated as indicative. A definitive ID would normally require scientific study of samples and in some cases genetic analysis.

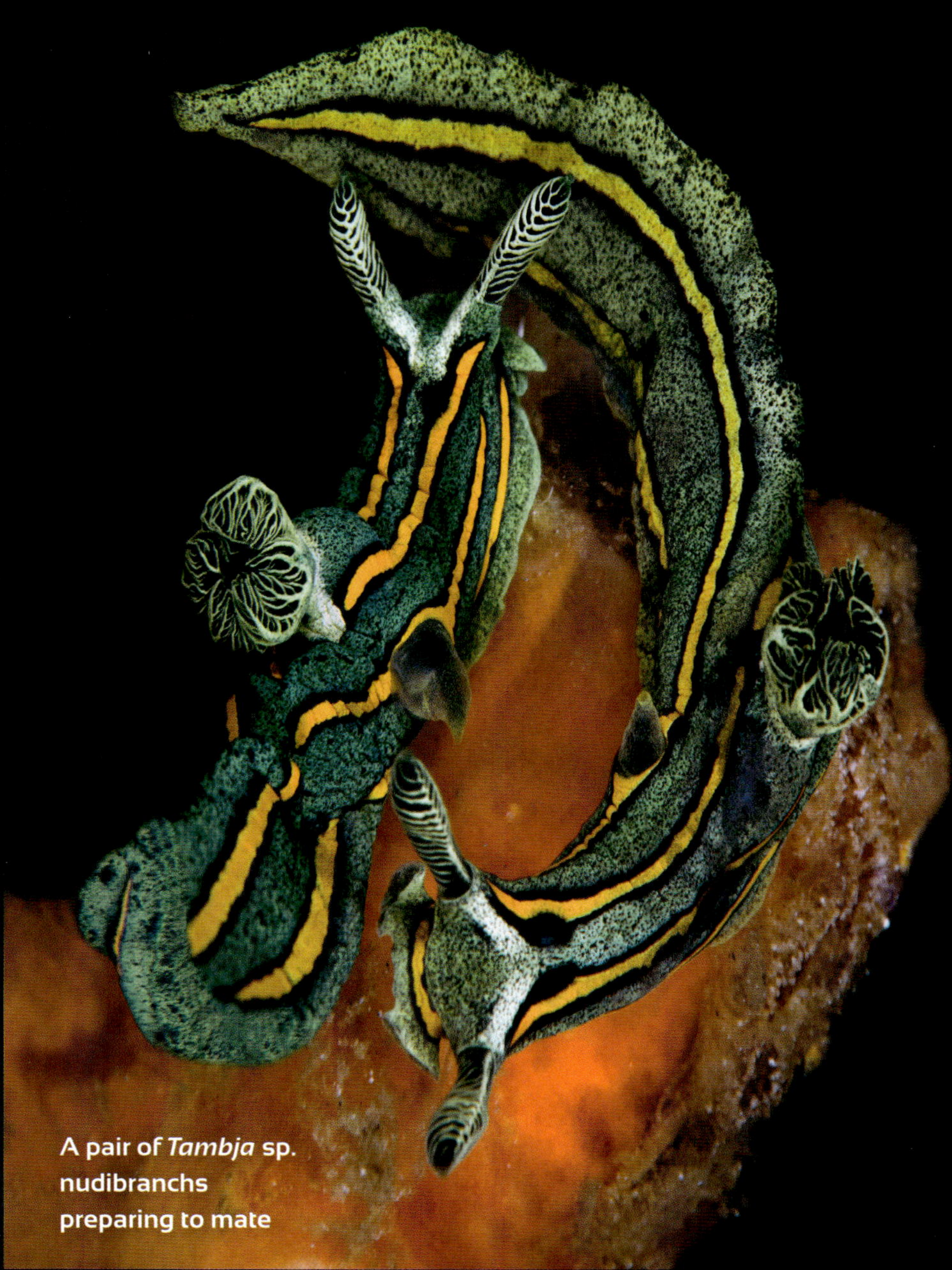

A pair of *Tambja* sp. nudibranchs preparing to mate

Foreword

About forty-five years ago we hovered over the first coral reef we had ever seen in what today is called 'The Coral Triangle'. Snorkeling above a remote drop-off in northern New Guinea, we stared in amazement at a nearly vertical wall of coral that pulsed with millions of small reef fish. Back then we focused on color and movement and could hardly identify the fish much less understand complexities of coral reefs. We did not know that a much closer look would take us beyond the multitude of forms and colors to an interdependent community of countless creatures; the hallmark of peak biodiversity on a healthy coral reef.

Alan Powderham and Sancia van der Meij's second collaboration, *Coral Triangle Cameos: Biodiversity and the Small Majority*, explains what our inexperienced eyes missed that seminal day in New Guinea. By revealing ways in which the smallest crustaceans feed among coral polyps, or how a preternatural octopus might use tools, or how certain molluscs derive energy from photosynthesis, the book highlights the connectivity of life on a coral reef. We learn that a honeycomb oyster's zigzag shell closure is not only beautiful, but also enhances water flow and thus nutrition for this filter feeder. And that brittlestars have the most species diversity of all echinoderms, with over eight hundred in the Coral Triangle alone.

Alan Powderham's photography and text brilliantly illustrates how reef creatures both depend and prey upon one another. The author and his scientific consultant, Sancia van der Meij, introduce us to reefscapes dominated by hard coral gardens and towering gorgonians. But as Powderham focuses his camera tightly with a macro lens we are shown glimpses of a hidden world. Corals not only harbor small, well-camouflaged fish and invertebrates, but in their role as a substrate, they also serve as eternal battlefields where the loser's remains are scavenged as soon as the conflict is over. With succinct prose, Powderham draws our attention away from the big picture to focus on the reef's minutiae: the endless variation of ocular development among reef inhabitants, or how some ascidians can regenerate a complete animal from just a tiny bit of themselves by employing special blood stem cells. And, who knew that ascidians are so biologically complex they are considered our closest invertebrate relatives?

In order to save these worlds within worlds, it is imperative to study them closely. This is where *Coral Triangle Cameos* excels. Shown in great detail, a coral reef's multitudinous niches, its almost mystifying array of forms, make sense in this dynamic environment. Powderham and Van der Meij's work supports E O Wilson's axiom that 'Organisms are all the more remarkable in combination.'

Burt Jones and Maurine Shimlock
Award winning photographers whose work can be found at www.SecretSeaVisions.com and are administrators of Indonesia's www.BirdsHeadSeascape.com website.

CHAPTER 1

CORALS

Sea fan,
Melithaea sp.

Feather star
(or crinoid)

Hard corals
reef panorama
Acropora sp.

Isopora palifera
Montipora sp.

Golden sweepers, *Parapriacanthus ransonneti*

Sea fan, *Melithaea* sp.

Coral reefs and Darwin's Paradox

The Coral Triangle, conjoining the Indian and Pacific Oceans, encompasses some six million square kilometres, extending east-west from Malaysia to the Solomon Islands, and north from the Philippines to Indonesia and Timor-Leste in the south. Boasting incredible variety, including seventy-five percent of the world's coral species and some 2,500 species of fish, it is considered to be the epicentre of marine biodiversity on the planet.

The first six pages of *Chapter 1* convey how spectacular its coral reefs can be and are a compelling introduction to Darwin's Paradox. This posed the following conundrum: What could explain the incredibly rich biodiversity of tropical coral reefs and how they could flourish in crystal-clear and apparently nutrient-starved seas? Of course a key factor was the essential nurture provided by the photosynthesis of microscopic algae embedded in the surface tissues of the coral and of which Darwin was quite unaware. Moreover, this symbiosis of the algae with their coral hosts, strongly relies on clear water allowing the sunlight to penetrate. Thus, somewhat ironically in this context, high water clarity is actually a key benefit. Even so, photosynthesis is not enough to provide the full answer and marine biologists have responded to this enduring challenge with additional proposals.

One such is the 'Island Mass Effect'. In this, the reef is said to develop its own physical geometry to directly control the circulation of the seawater within it. This in turn concentrates the plankton, thus enhancing the harvest of this vital first link in the vast and varied food chains of the reefs. Another theory focusses on the vast numbers of tiny reef fish which contribute up to forty percent of the overall diversity of reef fish — examples of which are featured in *Chapters 1* and *3*. They may appear in dense congregations, such as the anthias and sweepers, to form vibrant living tapestries. The far less obvious and aptly named cryptobenthic fish add a further substantial component to this biodiversity. They are key members of the so called 'small majority' and estimates indicate that cryptobenthic fish alone could produce up to sixty percent of the fish biomass present on coral reefs. A different hypothesis suggests that the solution lies within the concentrated flow through the cracks and crevices of the reef. There, super-efficient filter feeders such as sponges can operate undercover to extract vital nutrients. It has been calculated that such plankton filtering by sponges could contribute an amazing fifty percent of the overall metabolism of the entire reef. Yet another proposal focusses on the intrinsically rough surfaces of corals. With wave action at a microscopic level, these break down the boundary layers which normally form a shield between the corals and the surrounding ocean. This disruption enables the coral to access the exposed nutrients. With this and other creative proposals it would seem that the challenge to solve Darwin's Paradox will endure for a while.

The following four pages feature examples of large circular and spherical coral formations clearly thriving, beyond any paradox, between island masses and boundary layers.

Hard coral,
Acropora sp.
Golden sweepers,
Parapriacanthus ransonneti

Hard coral,
Acropora sp.

Hard coral,
Acropora sp.

Hard coral,
Leptoria phrygia

Hard coral,
Turbinaria sp.

Hard coral, *Fungia fungites*

Mushroom club

The Fungiidae family — known as mushroom corals — are predominantly solitary single polyp corals which may have multiple mouths, as shown in the elongated example opposite. Like other Cnidaria their tentacles are armed with stinging cells, called nematocysts. The arms of the polyps can be seen emerging to feed in the lower *Cycloseris* photo opposite and even more prominently in the later section 'Night and day'. As hexacorals, they form limestone structures to house their polyps. Their arms form a radial symmetry of six, though this geometry is not readily obvious. By comparison, the arrangement of the eight tentacled polyps of octocorals is generally quite plain to see. Individual polyps of mushroom corals reach impressively large sizes with some solitary species exceeding more than 25 cm. While the juveniles start life anchored to the substrate by a stalk, most species are free-living as adults. I have occasionally seen them collect in large aggregations, a mobility which is quite at odds with our normal conception of corals' sedentary lifestyle. They are also able to right themselves if turned upside down, for example by wave action. However, beyond all the biology, a prime attraction is their wonderful abstract patterns and intricate structures. Examples of abstract art forms of other corals are shown in the following six pages.

Cycloseris sp.

Cycloseris sp.

Pleuractis sp.

Cycloseris sp.

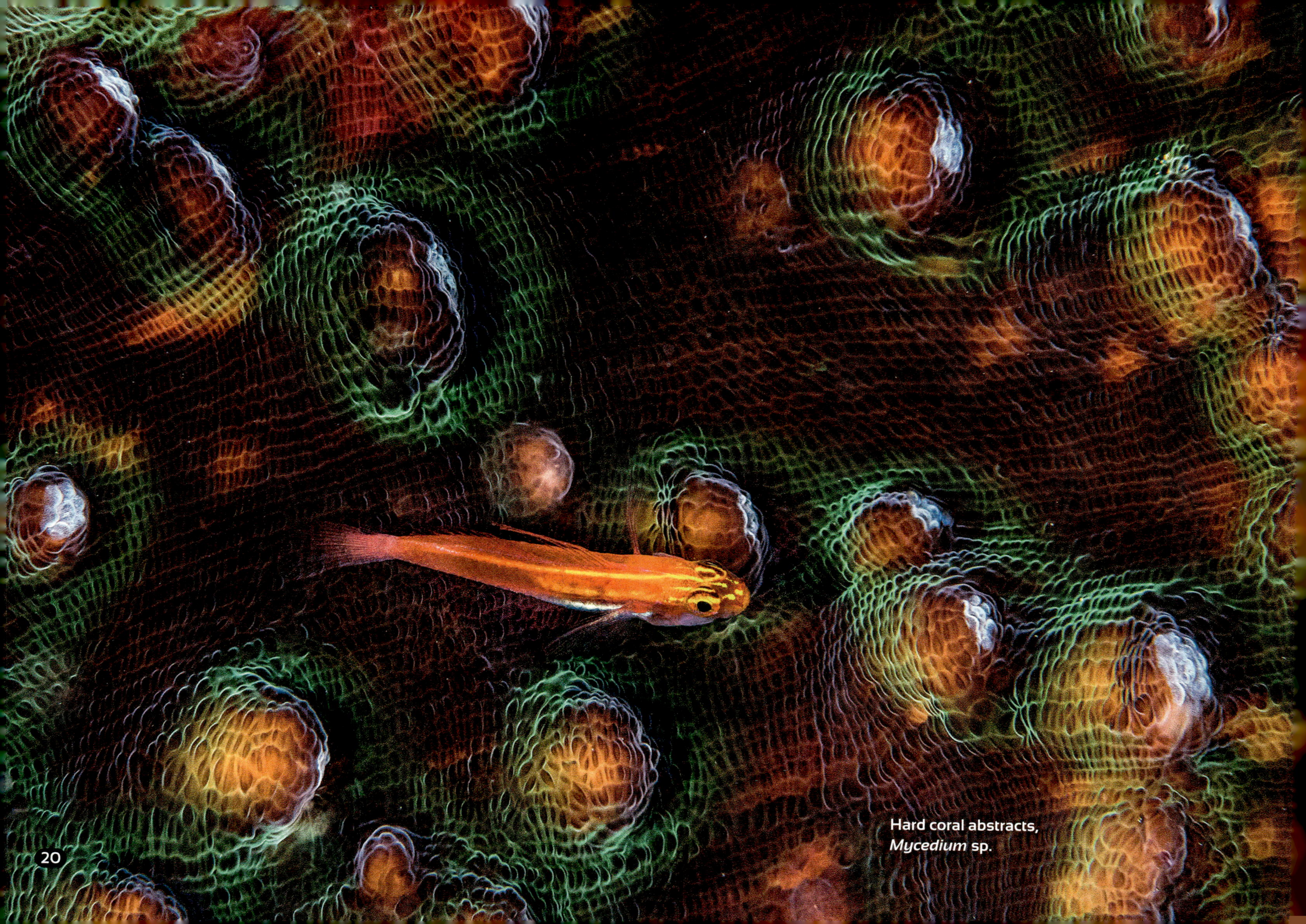

Hard coral abstracts, *Mycedium* sp.

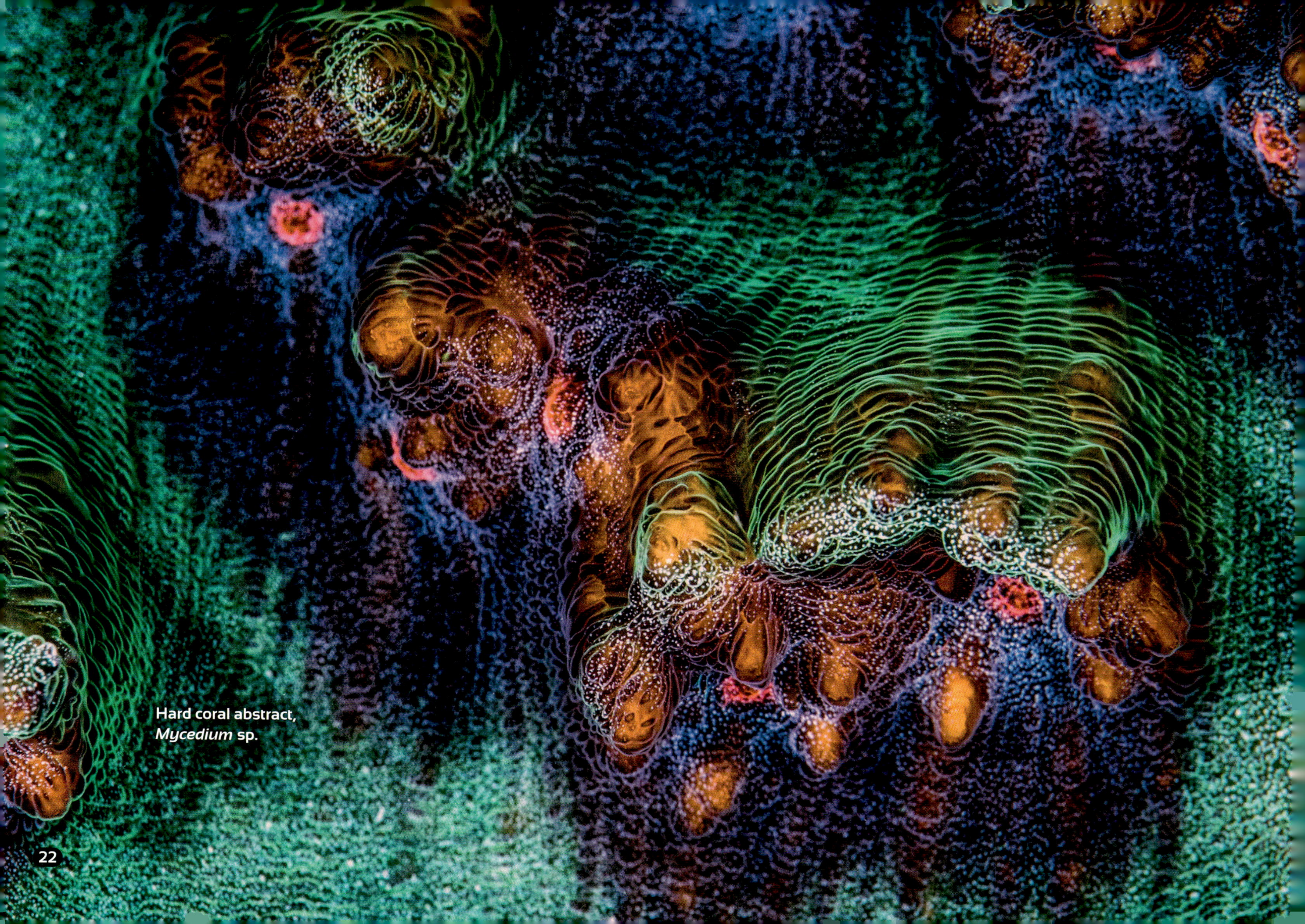

Hard coral abstract,
Mycedium sp.

Hard coral abstract,
Trachyphyllia geoffroyi

Hard coral abstract,
Dipsastraea cf. *favus*

Hard coral abstract,
Goniopora sp.

Hard coral, Fungiidae sp.

Night and day

The coral world features two main subclasses — the hexacorals and the octocorals. This section focusses on the dramatic differences across the twenty-four-hour cycle of life on a coral reef. The entire coral reef community is perpetually engaged in the diurnal/nocturnal struggle for survival. Octocorals, such as sea fans, do not have limited opening hours as their polyps stay extended unless physically disturbed. They also provide a relatively secure sanctuary for a multitude of associate species including small fish and invertebrates such as decapods. These may thus safely emerge at night from their diurnal hideaways to fulfil the vital needs of feeding and reproduction. Hexacorals, on the other hand, enjoy a night life very distinctive to their quiescent daytime disposition. At night, their polyps, which are kept retracted for protection during the day, extend to feed on the bounty of planktonic life carried in the upwelling currents. The photos here highlight such changes. The transformation of the bubble coral shown at night below illustrates how the bubbles, which are tightly bunched together during the day, are retracted to allow the polyp arms room to emerge. The *Favites* coral also shows the impressive reach of their polyps which, in some aggressive species, may be extended to attack and consume adjacent corals.

Bubble coral, *Plerogyra sinuosa*

Hard coral abstract, *Favites* sp.

Hard coral abstract, *Platygyra* sp.

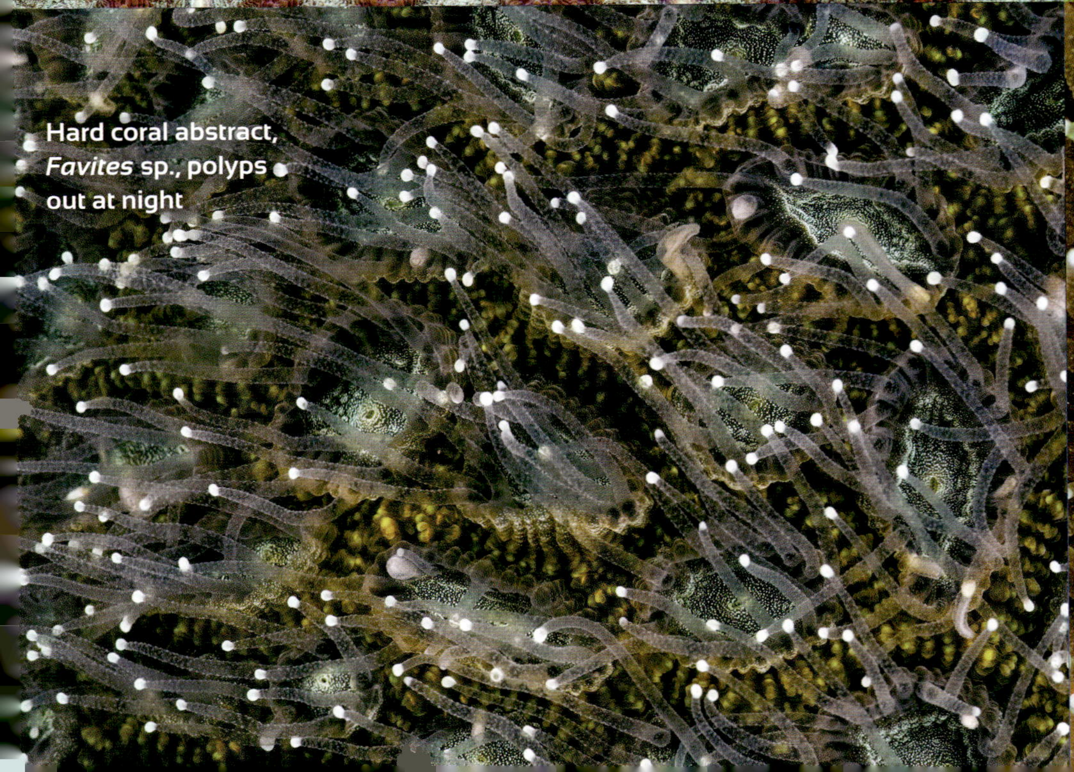

Hard coral abstract, *Favites* sp., polyps out at night

Hard coral abstract, *Platygyra* sp., polyps out at night

Rhythms of the reef

Previous sections have focussed on the hexacorals as these constitute the main reef-building corals. Through their ability to form calcium carbonate to house a countless myriad of tiny polyps they create the prodigious structural foundations for the entire biodiversity of coral reefs. If one is tempted towards the oft quoted term 'coral gardens' to describe these colourfully complex ecosystems, then it is the octocorals with their swaying sea fans that provide the illusion of exotic plants. But the pure wonder of coral reefs far exceeds our terrestrial endeavours to create the artifice of a country garden. Rather these incredible marine ecosystems are the completely natural result of evolution in harness with the tireless activity of countless tiny animals — the coral polyps.

While sea fans may provide a ready refuge, they inevitably attract predators such as the two species of hawkfish featured here. Being relatively small predators themselves, they receive a double benefit. The longnose hawkfish is nicely patterned to blend into the sea fan and is partial to pygmy seahorses (see the example overleaf). Thus, while finding their own prey, they simultaneously gain protection from larger predators. As both fish may associate with the same species of fan, it is more productive to seek the diminutive seahorses in fans where no hawkfish are resident. Any new arrivals of the former risk a short stay.

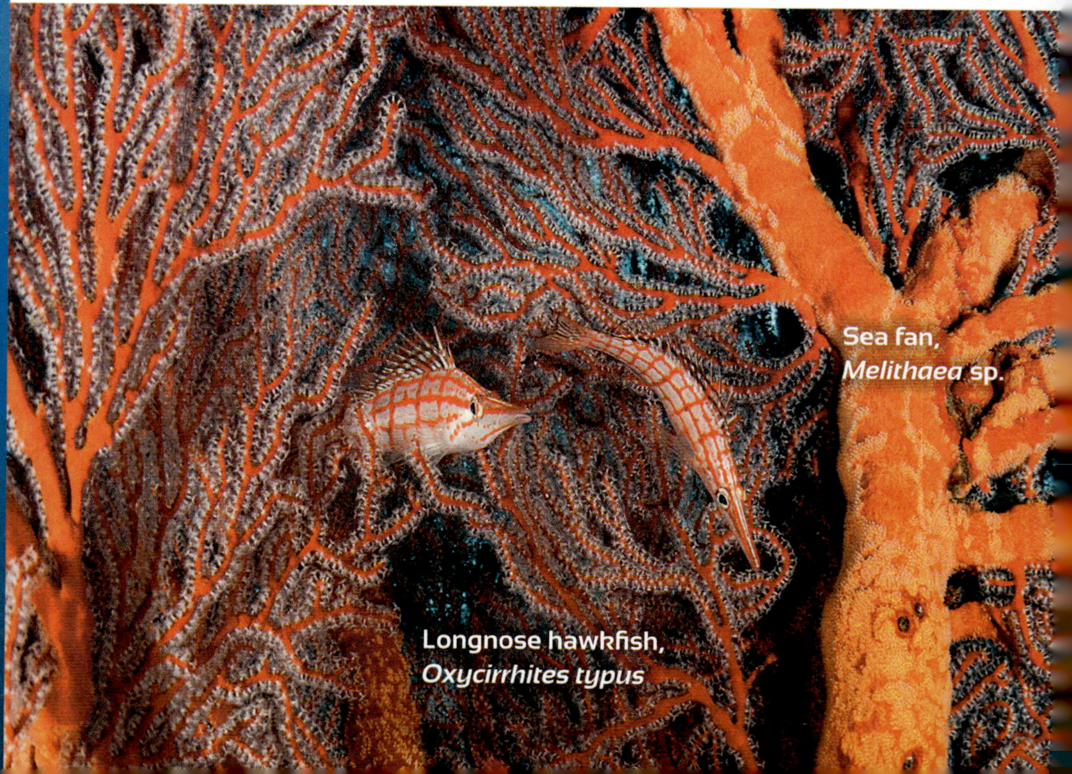

Octocoral polyps of organ coral, *Tubipora musica*

Hard coral, *Tubastraea micranthus*

Threadfin hawkfish, *Cirrhitichthys aprinus*

Shoal of silversides, *Atherinomorus* sp. in sea fan, *Melithaea* sp.

Red coral grouper, *Cephalopholis miniata*

Bicolor blenny, *Ecsenius bicolor* on *Leptoseris* sp. hard coral

Seeking sanctuary

Coral reefs, so fascinating in their own right, also offer refuge for a vast range of other creatures including, of course, the 'small majority'. Many species of tiny fish, such as the typically less than 4 cm-long cryptobenthic blennies, feature high metabolisms and fast rates of reproduction. Moreover, as they stay very close to the reef to avoid the risks of open water, their larvae do not widely disperse but settle locally. As previously discussed, estimates indicate that such species could contribute more than half of the fish biomass devoured on coral reefs. Such conspicuous consumption could therefore provide part of the answer to Darwin's Paradox.

Corals, in their many forms, provide sanctuaries, not only for inherently tiny fish but also for the juveniles of larger species. Three of the photos opposite featuring juveniles of damsel, file and cardinal fishes were taken at night. The former pair also display examples of the disruptive shading adopted at night for extra protection. The pygmy seahorse was encountered during a day with heavy cloud cover and at a depth of some twenty metres — hence the very dark background. Perhaps, as a prelude to imminent spawning, the polyp above the seahorse is seen releasing an egg within what appears to be a cloud of sperm.

Neon triplefin, *Helcogramma striata* on *Montipora* sp. hard coral

Neon triplefin, *Helcogramma striata* on sea fan, *Melithaea* sp.

Pygmy seahorse, *Hippocampus bargibanti* in soft coral, *Muricella* sp.

Cardinalfish, *Apogon imberbis*, in sea pen

Golden damselfish, *Amblyglyphidodon aureus* in soft coral, *Dendronephthya* sp.

Taylor's inflator filefish, *Brachaluteres taylori*, in soft coral, *Dendronephthya* sp.

CHAPTER 2

ANEMONES

Sea anemone,
Radianthus magnifica

A classic symbiosis

Symbiosis simply means 'living together'. However, behind this seemingly anodyne phrase lies an intricate web of interdependent relationships between the immense range of creatures that populate coral reefs. These relationships are fundamental to the health of these intricate ecosystems. There are three broad categories of symbiosis — mutualism, commensalism, and parasitism. The relationship between anemonefish and anemones has gained iconic status and, thanks to the film *Finding Nemo*, is now the most familiar example. However, the reality behind this apparently charming arrangement is curiously dark and complex. It is mutualistic and so one in which both parties benefit but, in this case, is hardly one of equality. For a start, the fish are obligate partners meaning that they cannot survive separately from the anemone. Yet this is with a partner that has lethal stinging cells and will consume small fish. To intimately nestle among those threatening arms, the anemonefish has clearly evolved a keen art in risk management. How this acclimatisation is safely achieved and maintained is still not fully established but recent research has identified a special protein that provides anemones with the ability to decide when and what they sting. There are hundreds of species of anemones, but only ten are symbiotic with the thirty-one species of anemonefish. The relationship has thus evolved to be more optional and selective for the anemones.

Anemonefish,
Amphiprion clarkii

Pink anemonefish (with juvenile), *Amphiprion perideraion*

Radianthus magnifica

Orange-fin anemonefish, *Amphiprion chrysopterus*

Sea anemone, *Entacmaea* sp.

Sea anemone,
Entacmaea sp.
Spinecheek
anemonefish male
Spinecheek
anemonefish female,
Premnas biaculeatus

Sea anemone (tentacles retracted), *Radianthus magnifica* with pink anemonefish, *Amphiprion perideraion*

Sea anemone, *Radianthus magnifica* with pink anemone-fish and associate shrimp

Pair of squat shrimp, *Thor amboinensis* on associate anemone

Obliging hosts

The previous pages provide various examples of the now famous symbiosis between anemonefish and their anemone hosts. It is a surprisingly close relationship which is especially evident during the periods when the anemone fully withdraws its tentacles within its mantle. This presents the anemonefish with two options — either to hover exposed on the surface of the mantle or, alternatively for better protection, to take the plunge by diving into the disappearing fronds of stinging arms. As can be seen, in the range of closed and partially closed anemones, the latter tactic is the favoured choice. Quite what factors cause the anemones to withdraw into these closed phases is not fully understood, but such action clearly creates extra drama in the already colourful life of anemonefish.

The resilience of anemones is highlighted later in this chapter and in fact they seem to thrive in the most exposed areas of the reef. They are tough characters and quite adept at making their own dramatic statements. The vivid crimson mantle, so brightly displayed by the closed anemone on the previous page, makes for a surreal encounter — especially with it set in splendid isolation on top of an otherwise bare reef. The anemonefish look less than relaxed even when escorted by a parade of cleaner shrimp. As shown here, anemones also provide sanctuary for various other species of decapods thus extending their symbiotic repertoire.

Porcelain crab, *Neopetrolisthes maculatus* on anemone, *Stichodactyla mertensii*

Anemone shrimp, *Ancylocaris brevicarpalis*

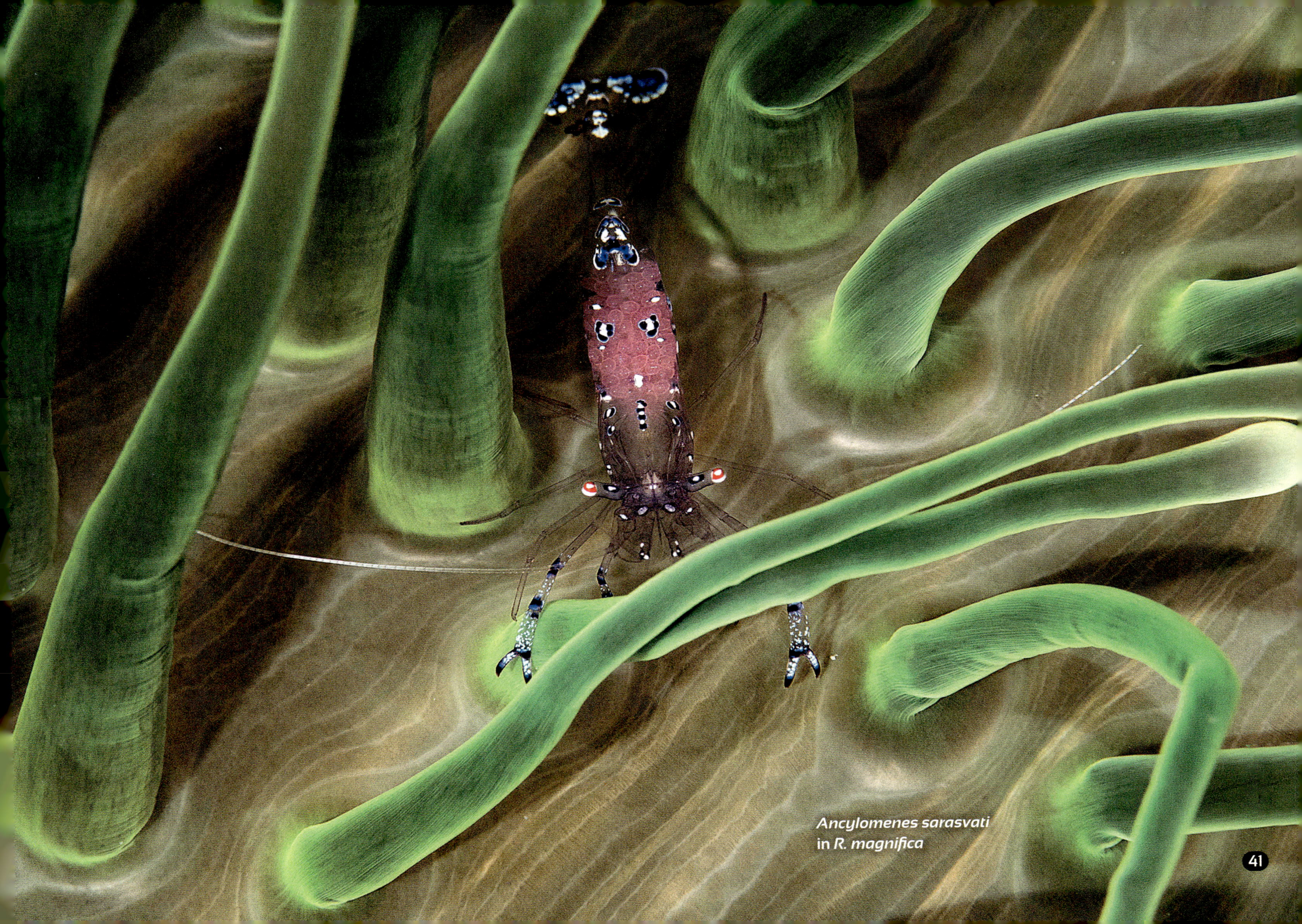

Ancylomenes sarasvati
in R. magnifica

Anemonefish, *Amphiprion clarkii*, with anthias and sponges sharing reef dominated by sea anemones

Medusal flair and Greek myths

When fully open, tube anemones display a flowing pallet of vivid colours. They seem to glow as if internally lit with deep pinks, neon greens, pulsating blues and vibrant violets. The word anemone derives from the Greek for a species of wildflower which translates to 'windflower'—a reasonable etymology for the marine examples here. Certainly, as these captivating creatures wave in the currents, the term appears most apt. It is tempting to extend the analogy further since, in Greek legend, these flowers grew from the tears of Aphrodite upon the death of her lover Adonis. They are rich in symbolism having specific meanings relating to different colours. And the colours here certainly resonate. However, this may be stretching the connection a bit far since these are not true anemones. While sharing the same phylum of Cnidaria, tube anemones belong to quite a different class known as Ceriantharia. As shown, they have a distinctly different arrangement of tentacles with short ones surrounding the mouth and an outer ring of much longer ones which capture their planktonic prey. But the allusion is not entirely lost as *Cerianthus* in Greek means 'wax flower'. Tube anemones have no symbiosis with anemonefish but various species of shrimp and crab may be seen enjoying protection on the upper part of the tube beneath those swaying tentacles.

Tube anemones,
Cerianthus sp.

Tiger anemone, *Nemanthus annamensis*

Gift-wrapped gorgonians

The evocatively named zebra-striped gorgonian wrapper is a colonial parasitic species native to Indo-Pacific waters. The leopard shrimp is its symbiont whose striped patterns echo those of the anemone making it super-cryptic and a real challenge to spot. So, in this case, the leopard has lost its spots. The gift in the title is entirely to the benefit of the anemone. It literally wraps itself around the arms of gorgonian sea fans and whip corals, subjecting them to a slow and creeping demise as they are progressively overgrown by these invasive colonialists. In a deadly example of a prime real estate land-grab, the corals are reduced to a skeletal framework on which the anemone establishes its own colony. They are thus perfectly situated in the current to feed upon a regular stream of plankton. The anemones have two ways of initiating their invasions. One is through the typical process of larval dispersion and settlement common to the life cycle of many marine fauna. But these anemones also have a more mature approach, as individual polyps have the optional ability to detach and so spread pelagically to new locations. Each polyp has upwards of 120 tentacles, with those of the inner ring being significantly longer that the outer ones. This, interestingly, is exactly the reverse arrangement of the tube anemones previously described. Evolution follows many paths.

Tiger anemone with associate shrimp, *Izucaris masudai*

Anemone tenacity

Superficially, sea anemones may appear to be somewhat fragile. Those flowing fronds of tentacles and soft cushion-like mantles tend to enhance this perception. This is a serious misconception since anemones are robust and prolific carnivores. Their tentacles are armed with arrays of harpoon-like stinging cells which inject powerful neurotoxins enabling anemones to stun and devour relatively large prey. At the more extreme end of their cuisine, I have encountered one engulfing a complete flounder and another consuming a large sea star. But beyond dietary options, the choice of location strongly undermines any concept of fragility. Those tentacles have abrasive surfaces and the anemones can very effectively fend off the encroachment of both hard and soft corals. As pictured here and overleaf, these large anemones have adopted some of the most exposed sites on the reef, sitting high at the crest. Here they are continuously buffeted by strong currents, making their abundance of long tentacles wave dramatically in furrows like a field of wheat in a storm. Yet it is the anemonefish that seem challenged — they certainly get plenty of swimming exercise. Strong currents provide a rich bounty of plankton and recent studies have shown that sea anemones grow more tentacles when well fed. The fecundity and robustness of anemones is also evident where they proliferate to form vast fields that can dominate a reef as shown in the last two pages of this section.

Gorgonian,
Isis hippuris
Sea anemone,
R. magnifica
Soft coral,
Sinularia sp.

Sea anemone, *R. magnifica* with anemonefish, *Amphiprion ocellaris*

Sea anemone, *R. magnifica* with bleached tentacles and pink anemonefish, *Amphiprion perideraion*

CHAPTER 3

FISH

Blackfin barracuda, *Sphyraena qenie*

Hard coral, *Acropora* sp. with clouds of feeding anthias

Mixed shoal of anthias enveloping green hard coral, *Tubastraea micranthus*

Safety in numbers

A rich abundance of small fish populate a healthy reef. The spectacular hordes of anthias are instantly obvious but there are also more secretive 'small majorities'. These latter include the aptly named cryptobenthic fish which have been estimated to constitute more than half of the total biomass of reef fish. They all depend on the ready supply of plankton which forms the first key link in the entire food chain. Diurnal feeding fish, such as the anthias pictured left, rise into open water above their coral refuges to feed. Here they must brave the risk of predation from larger fish such as jacks, groupers and, as featured on the chapter page, barracudas. The aggregations of sweepers shown opposite, so dense that they might merit a description of 'packed in like sardines' shelter beneath an Acropora coral. Here, they rest in their daily dormant mode. They emerge to feed at night but then must face the onslaught of nocturnal predators such as soldierfish. From a purely artistic perspective, these dazzling arrays of colourful fish present great photographic potential. But to capture such vivid images with an adequate depth of field, demands a fisheye lens placed in close proximity to the subject. The visually intrusive and noisy expirations from conventional diving gear will promptly disrupt any tranquillity. Here the discrete silence of a rebreather is paramount and many of the close encounters featured in this book owe their success to the effectiveness of this equipment.

Shelving stands of hard coral, *Acropora* sp.

Golden sweepers, *Parapriacanthus ransonneti*, sheltering below table of *Acropora* sp.

Sea fan, *Annella* sp.

Red lionfish, *Pterois volitans*

Fantastic beasts

The smaller but vital denizens of coral reefs must run the gauntlet of open water which is constantly patrolled by hordes of fast-swimming piscivores, but they also face sudden attack from below such as from the dwarf scorpionfish pictured opposite. The lines of bright pigment which help to disguise the eyes are supplemented by the garish yellow patterns which act as disruptive camouflage. The red lionfish on the left, operate singly or in small groups. While they will happily engage in a sedentary waiting game, they also employ more active tactics by using their intimidating array of fins and venomous spines to corral and trap their victims. Consequently, they enjoy a wide spatial domain and will hunt at all depths from surface to substrate. Other members of this lethal band of voracious predators are featured on the following four pages. They are masters of creative confusion, by blending into the background as exemplified by the reef scorpionfish, or have very disruptive colours and patterns including 'see through' effects such as the fin windows of the *Rhinopias* or the filigree filaments of the hairy frogfish. The latter adds the deployment of a lure to tempt unwary prey which, if venturing too near, face the fastest known strike of any vertebrate. The purple coloured *Rhinopias* has quirkily departed its usual benthic domain to perch in a gorgonian. These exotic predators bring abstract art to ambush.

Ragged-finned firefish, *Pterois antennata*, detail

Dwarf scorpionfish,
Scorpaena elachys

Scorpionfish,
Scorpaenopsis sp.

Weedy scorpionfish,
Rhinopias frondosa

Weedy scorpionfish, *Rhinopias frondosa*

Hairy frogfish, *Antennarius striatus*

Whale sharks, *Rhincodon typus* with remoras, Echeneidae sp.

Marauding majorities

While stretching the concept of 'small majorities,' these somewhat bizarre examples do raise interesting questions about behaviour and gender. The encounter with the whale sharks involved two males and two females, one of which was dramatically dressed in a shroud of over eighty remoras. It is common to see small numbers of remoras acting in symbiosis as commensals with large pelagic fish. Such an impressively huge aggregation like this, while not unknown, is rather curious. The other three whale sharks had only a very modest attendance of remoras. It is natural to wonder what made the particular female so attractive to so huge a mob of remoras. The example opposite also features unusual behaviour. The symbiosis between anemonefish and anemones is recognised as the gold standard for such relationships in the Indo-Pacific. As their name suggests, Banggai cardinalfish are naturally endemic to the Banggai archipelago in Central Sulawesi where they do indeed associate with anemones. But this photo, showing the cardinalfish surging over a cowering anemonefish, was taken in a quite different environment over five hundred miles away in Lembeh. There is a sad irony here since, while these isolated populations proliferated from aquarium trade escapees, the Banggai in their natural home are now an endangered species. Both of these cases are considered in more detail in *At the Heart of the Coral Triangle* (2021).

Banggai cardinalfish, *Pterapogon kauderni* swarm over pink anemonefish and closed anemone

Blunthead batfish, *Platax teira*

Dawn chorus

At the advent of scuba diving, Jacques Cousteau famously coined the phrase 'The silent world' to convey the other-worldly experience of diving in this new frontier. The description still enjoys some currency today but in reality coral reefs are far from silent. In fact they are quite noisy environments. Fish create cacophonies of vocalisations, from chirping anemonefish to grunting sweetlips, all in concert with the continuous rasping of parrotfish and other grazers scraping away at the coral. Batfish gather over the reef at dawn and along with other species have been recorded performing in a dawn chorus rather like birds. In fact, recent studies indicate that corals actually prefer noisy neighbourhoods and around fifty percent more coral larvae have been found to settle on reefs with a healthy fish population. However, these natural sources of sound are largely obscured by the noisy exhalations of bubbles from standard scuba equipment. Thus, full appreciation of this sound-world is largely veiled from the typical diver. At best such a diver can only get a hint of it between breaths which rather diminishes the experience. To be fully immersed in the quasi-silence and thus appreciate the intrinsic concerts of the reef, one needs a rebreather. They also enable a discrete intimacy for close observation of reef life and stimulate a variety of fish, from blennies to sharks, to indulge their natural curiosity. The quiet observer when fully accepted, thus becomes the observed.

Pinnate batfish, *Platax pinnatus*

Ribbon sweetlips, *Plectorhinchus polytaenia*

Blunthead batfish, *Platax teira*, juvenile

Beguiling batfish

As already noted, adult batfish boldly celebrate the freedom of open water with the equivalent of a piscine dawn chorus, making no attempt at disguise or concealment. As juveniles, however, they indulge in a fascinating range of deceptive behaviours. They are masters of creative deception which they deploy either at the very surface or down deep on the substrate. The round and blunthead juveniles specialise in surface impressions where they imitate both the pastel palette and action of drifting debris. The round are especially adept at this skill, effectively hiding in plain sight as they mimic floating dead leaves. The pair pictured opposite accompany such a leaf, the upper three images in the photo being simply reflections in the water surface. Far below on the substrate, the juveniles of another two species of batfish adopt different arts of disguise. The Batavia below, typically associates with black and white feather stars and its intricately striped livery enables it to subtly blend in as it nestles among the arms of the sea star. Heavy dark bands complete this disappearing act by concealing the eyes. Longfin batfish have evolved a radically different strategy. Instead of blending in, their juveniles overtly stand out in a Batesian mimicry by echoing the bright aposematic warning colours and motion of a foul-tasting flatworm. The younger juveniles will enhance the deception by lying flat and swimming sideways.

Zebra batfish, *Platax batavianus*, juvenile

Pinnate batfish, *Platax pinnatus*, juvenile

Round batfish, *Platax orbicularis*, juvenile

Cryptobenthic charmers

Cryptobenthic, which basically means 'hidden on the bottom', refers to the tiny, secretive fish which typically hide within the coral crevices and reef debris. Consequently, they are easily overlooked but are now, as noted in the section on 'Coral reefs and Darwin's Paradox,' considered to be key contributors to the very viability of the reefs. Recent research, fuelled by the catalyst of resolving this paradox, has focussed on discovering the apparently huge deficit in the nutrients required to sustain the complex ecosystems of coral reefs. The larvae of cryptobenthic fish are not broadcast but settle locally thus creating a bounteous cycle of nutrition concentrated within the confines of the reef. Blennies are prominent citizens of this community and are widely represented in tropical and temperate seas with an estimated 350 species. They provide a variety of charming characters as they peer out from their hideaways with large appealing eyes. This description is qualified since a sprinkling of the notorious fang-blennies number among this family. Most blennies are herbivores but the fang-blenny is the black sheep exception. It operates in borrowed clothes as it mimics cleaner fish. This enables it to approach and take a bite out of unsuspecting patrons attending cleaning stations. A series of bona-fide cleaners, including shrimps, servicing their clients is provided on the following pages along with a variety of other small reef fish.

Tail-spot blenny, *Ecsenius stigmatura* in hard coral retreat

Bluestriped fang-blenny, *Plagiotremus rhinorhynchos*

Sabretooth blenny,
Petroscirtes xestus

Pygmy angelfish, *Centropyge bispinosa*

Bignose unicornfish, *Naso vlamingii*

Threadfin anthias, *Pseudanthias huchtii*

Bluespotted cornetfish, *Fistularia commersonii*

Yellowtail fusilier, *Caesio cuning*

Emperor angelfish, *Pomacanthus imperator*

Regal angelfish, *Pygoplites diacanthus*

Red-toothed triggerfish, *Odonus niger*

Elegant dartfish, *Nemateleotris decora*

Fairy basslet, *Pseudanthias* sp.

Fire dartfish, *Nemateleotris magnifica*

Princess damselfish, *Pomacentrus vaiuli*

Threespot angelfish, *Apolemichthys trimaculatus*

Taylor's garden eel, *Heteroconger taylori*

Mandarinfish, *Synchiropus splendidus*

Longnose butterflyfish, *Forcipiger flavissimus*

CHAPTER 4

CEPHALOPODS

Broadclub cuttlefish,
Ascarosepion latimanus

Flamboyant variations

Even among the extrovert cephalopods, Flamboyant cuttlefish stand out as true eccentrics. That is quite an achievement given the diverse range of colours, shapes, textures and patterns that cuttlefish have at their disposal in their wondrous wardrobe. A flavour of this versatility is shown in the examples overleaf. The larger species of cuttlefish, such as the broadclub pictured on the chapter page, will confidently patrol above the reef in open water. Small cuttlefish, however, are high on the menu of many predators and so naturally seek refuge within the reef. Here they deploy their consummate abilities in disguise to blend in. Although the flamboyant cuttlefish shares a comparable range of such skills it frequently favours a radical and seemingly perverse approach. Certainly, as shown top left, it can render itself virtually invisible with a convincing imitation of a drab sandy substrate. However, its default mode is to abandon this deception and blithely parade across the sand in the most garish of livery as shown opposite. Compared to its cousins, this 8 cm extrovert is a poor swimmer and will only occasionally rise into the water column as pictured below. In its preferred benthic lifestyle, it employs a curious mode of ambulation. Using its lower arms and the rear parts of its mantle it boldly walks over its domain. Such confidence derives from having the strongest toxins among cuttlefish. Its bright aposematic colouration dramatically displays the warning threat of this potency.

Flamboyant cuttlefish,
Ascarosepion pfefferi

Reaper cuttlefish, *Ascarosepion* cf. *mestus*

Face to face

Reef cuttlefish are generally encountered singly but, as shown left, will form pairs as a prelude to mating. The one in front is prominently displaying a V-sign. It is a common gesture among cuttlefish though, given the circumstances, it is probably not intended to convey a message of greeting to a diver. Cuttlefish may be observed during the day practicing their ritual courtships. They can be quite extended affairs in which they hunt and dine together before finally joining face-to-face in a writhing embrace of their combined sixteen arms. They celebrate the excitement of their couplings with waves of pulsating colours which flow across their bodies. These vivid displays are activated by the neuromuscular organs of their chromatophores. During the day these exhibitions are impressive but, when viewed nocturnally, the intensity can be simply mesmerising. Such psychedelic effects may be seen at their most heightened with close encounters of squid hunting at night. Some colourful examples of cuttlefish face makeup and disguise are shown in the gallery opposite. While, in this case, they have adopted a predominantly crimson mode to match their surroundings, this line-up conveys an engaging versatility. The upper pair convey wispy confections to match their backgrounds, while could one of those immediately below be mimicking a crab with huge red pincers? Only the cuttlefish really knows.

Ascarosepion cf. *mestus* mating

Reef cuttlefish, *Sepia* sp. colour variations

Squid game

Among invertebrates, cephalopods are widely admired for their intelligence. They match some of the smartest vertebrates which is pretty impressive for a mollusc. Yet, being molluscs, they have evolved along completely distinct pathways. So, rather like their eyes as noted in *Chapter 5*, their brains are cited as prime examples of convergent evolution. A direct experience of this mental prowess was provided early in my introduction to diving in the Coral Triangle. My interest then was predictably attracted by the more obviously impressive subjects such as dense shoals of large fish. Underwater, good photography generally demands close proximity to the subject — a familiarity not readily welcomed by the creatures of the reef. So I was delighted, when encountering a large shoal of barracuda, to find that my local dive guide was well versed in the right procedure. He immediately signalled me to avoid further approach and wait while he circled deeper to take position on the far side of the shoal. From this manoeuvre he was able to shepherd the fish towards me. So it came as quite a surprise when he blankly refused to apply the same tactics when we happened upon a school of squid. 'Squid, like cats — too smart — don't herd like barracuda' was his amusingly laconic but insightful explanation. Examples of squid night-time antics are provided on the following pages including the excavation skills of the tiny bobtail.

Bigfin reef squid,
Sepioteuthis lessoniana

Hummingbird bobtail squid, *Euprymna berryi*

Bigfin reef squid,
Sepioteuthis lessoniana,
rear view

Bigfin reef squid,
***Sepioteuthis lessoniana*,**
front view

Coconuts to green bottles

With their impressive adaptive DIY abilities, veined octopus were using shells as protective shelters long before we started dumping sundry refuse like bottles and cans into their environment. Such objects are but variable fare to octopuses' innovative skills in ad hoc home construction. But it was not until 2009 that a team of marine biologists, studying tool use by animals, reported the veined octopus's creative innovations with the discarded halves of coconut shells. It was seen as a quantum leap for an invertebrate. They were even observed to carry these trophies around by walking with them held between their arms. This is quite a feat in itself since they do not have legs like vertebrates but rely on hydraulics to achieve bipedal locomotion. If they can find matching coconut halves then combining them creates luxury accommodation. Such anticipation would imply that the octopus actually plans ahead. So this enterprising cephalopod has gained lasting fame as the coconut octopus. As shown, the octopus does not depend on perfectly matching halves but is fully adept at combining whatever material it encounters. A most impressive skill, placing it at the pinnacle of invertebrate accomplishment. Such behaviour has not yet been observed in other species of octopus. But then, if you are one of the most venomous marine creatures, like the blue-ringed octopus featured overleaf, there is no need for extra protection.

Veined or coconut octopus, *Amphioctopus marginatus*

Blue-ringed octopus,
Hapalochlaena sp.

CHAPTER 5

EYES

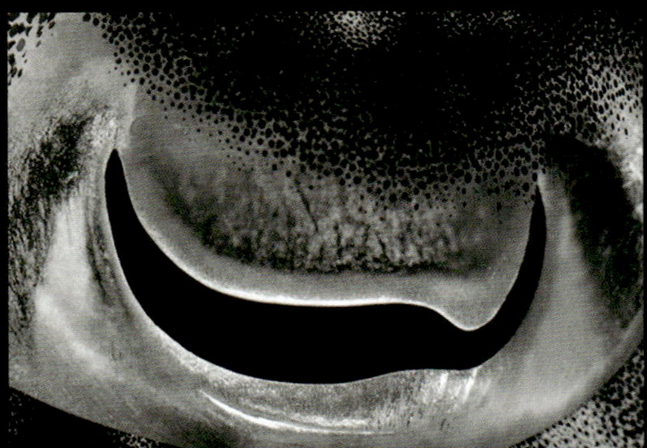

Crocodilefish,
Cymbacephalus beauforti

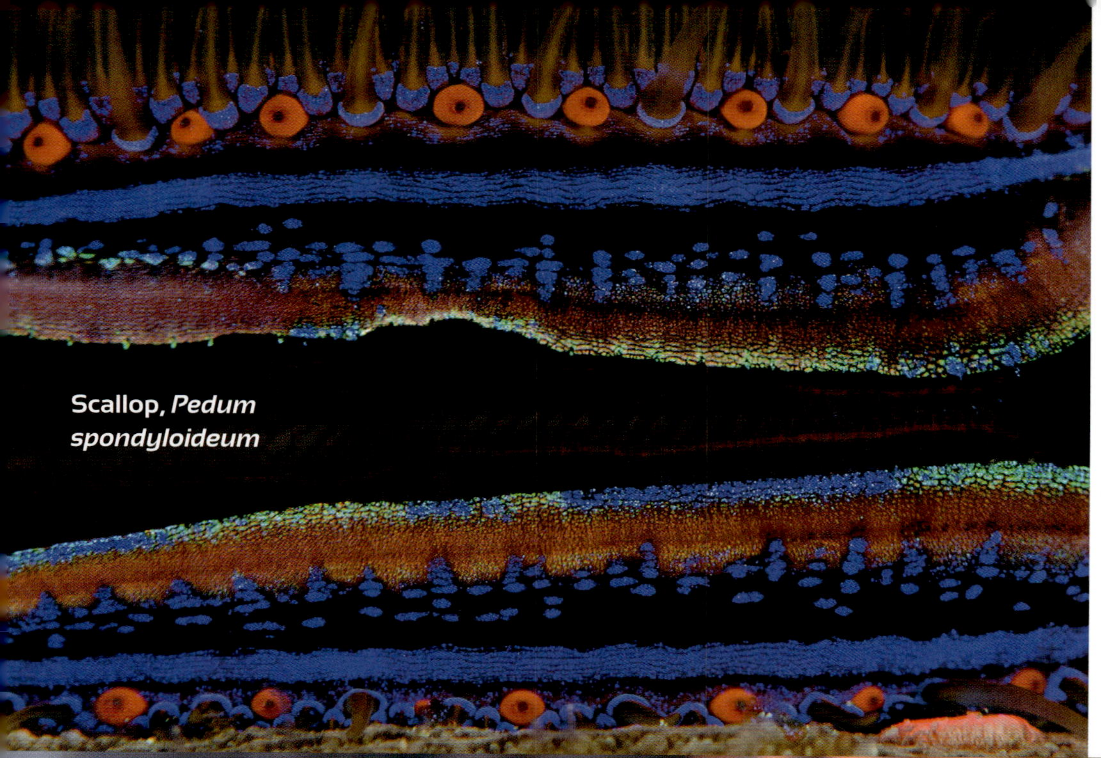

Scallop, *Pedum spondyloideum*

In the blink of an eye

Eyes have a lot to answer for. Their comparatively sudden arrival unleashed a brutally competitive world, drawing both hunters and hunted into an evolutionary arms race. Later, the apparent perfection of the eye fuelled another conflict between evolutionary biologists and creationists. Just how many times the eye has evolved has also been a hot and enduring debate. However, there is now broad agreement that the first eyes appeared with the Cambrian explosion around 543 million years ago. This is quite incredible since, if you wind back just a further million years — a mere blink in geological time — eyes are literally nowhere to be seen. A sample of the richness in ocular variation is shown here. Scallops are endowed with up to two hundred eyes — present in this example as the tiny red spheres lining the mantle margins — and they are far more complex than simple pin-hole cameras. Each eye has a concave mirror for focus and two retinas affording both close and peripheral views simultaneously. Mantis shrimps have the most complex of visual systems. Incredibly, each eye has some ten thousand photoreceptors which provide independent 3D vision in each eye and across many wavelengths. The other photos here and overleaf illustrate the robustness and utility of shrimp eyes. Despite any apparent fragility, the one below is seen bursting from concealment through abrasive sand and that opposite is adventurously negotiating the perils of a passage through open water.

Peacock mantis shrimp, *Odontodactylus scyllarus*

Tiger prawn, *Penaeus* sp.

Cleaner shrimp,
***Urocaridella* cf.**
pulchella

Humpback prawn,
Metapenaeopsis sp.

Anemone shrimp, *Ancylomenes sarasvati*

Cuttlefish, *Ascarosepion latimanus*

Exemplars of convergent evolution

Cephalopods, such as those pictured here, have impressively complex eyes. While they may appear strikingly similar to vertebrate eyes, they have evolved quite separately. In fact, cephalopod evolution has produced superior eyes. A key difference is that they feature everted retinas as opposed to the inverted form found in vertebrates — including humans. The former arrangement has the benefit of avoiding the presence of a blind spot which is caused by the alignment of the optic nerve passing through the retina and creating a small area without photoreceptors. Moreover, vertebrate eyes could be said to have evolved back to front (i.e. inverted). This is because, with the photoreceptors located at the back of the eye, the light rays are slightly diminished by having to pass through all the preceding layers. In cephalopods, the photoreceptors are at the front and so do not suffer this reduction in light intensity. Other variations include a lens which is spherical, whereas ours are ellipsoid, and in cephalopods focus is achieved like a camera by moving the lens — as opposed to changing the lens profile as we do. The W-shaped pupil of the cuttlefish affords them excellent all-round vision and a similar benefit is evident in the photos of the bobtail squid. Intriguingly, could the mimic octopus pictured opposite perhaps be imitating a pair of huge eyes in a defence display?

Veined octopus, *Amphioctopus marginatus*

Mimic octopus, *Thaumoctopus mimicus*

Hummingbird bobtail squid, *Euprymna berryi*

Bobtail squid, rear view

Bigfin reef squid,
Sepioteuthis lessoniana

Sepioteuthis lessoniana,
eye detail

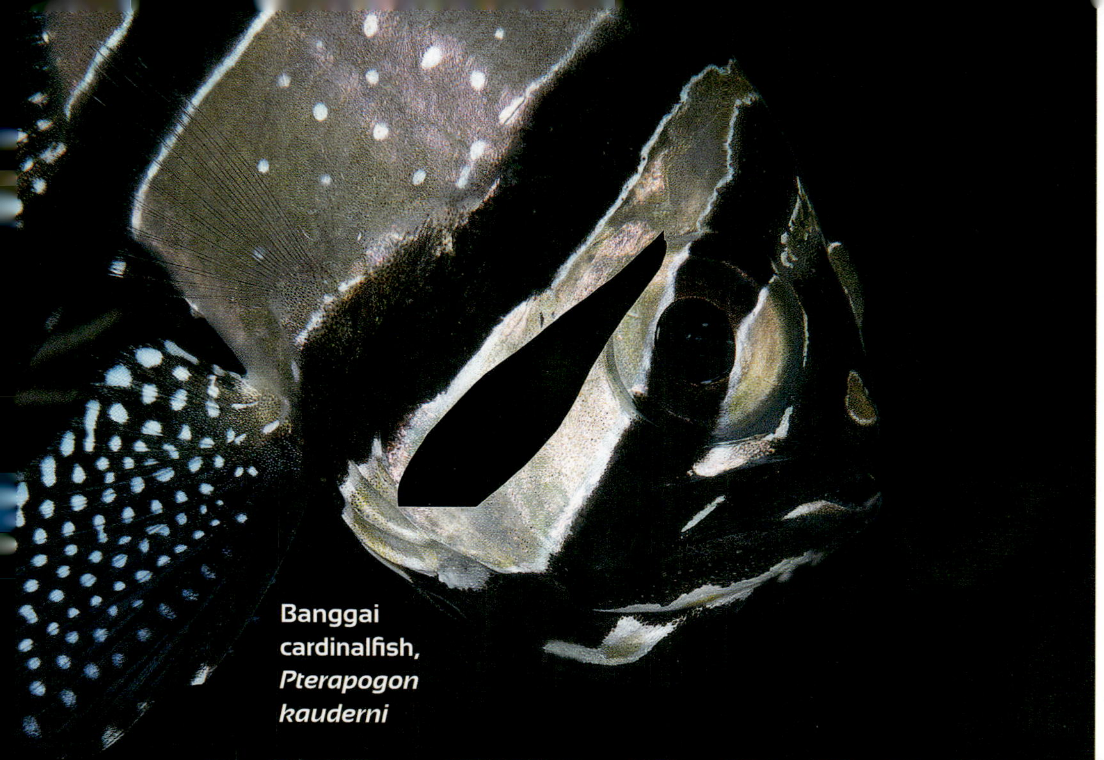

Banggai cardinalfish, *Pterapogon kauderni*

Eye lines and low brows

Eyes are one of evolution's outstanding inventions, especially given the bewildering range of designs. The eyes of many small fish are obscured with prominent lines such as featured by cardinalfish, pufferfish and tangs. The latter also include a strong colour signal at the tail to divert an attack. Disruptive eye makeup is also exploited by predators. Compelling examples are found in scorpionfish (see 'Fantastic beasts' in *Chapter 3*) and illustrated below by the *Rhinopias* whose eyes are lost in a confusing matrix of patterns and colours. Bony fish do not have eyelids but in this class, puffers have the unique ability to close their eyes by withdrawing their eyeballs deeply into their sockets by some seventy percent of the eye's diameter. This combined with the capacity to pucker the adjacent skin, enables them to manage a passable impression of blinking. This visual mobility allows them to adjust focus over a range of distances. The porcupinefish adds a glittering presence to this range of skills with the ultimate eye makeup of iridescent pigments. These help to shade their eyes in the manner of Venetian blinds. Flounders and soles have evolved an extreme example of eye mobility. As they develop from larva to juveniles, one eye literally migrates through the skull, displacing bone, to join the other — thus enabling these ambush predators full visual acuity while lying flat. The hemispherical lenses as featured by the grouper and soldierfish attest to the importance of all round vision.

Palette surgeonfish, *Paracanthurus hepatus*

Weedy scorpionfish, *Rhinopias frondosa*

Map pufferfish, *Arothron mappa*

Peacock flounder, *Bothus lunatus*

Shortspine porcupinefish, *Cyclichthys orbicularis*

Peacock sole, *Pardachirus pavoninus*, with nostril showing below eyes

Black-tipped grouper, *Epinephelus fasciatus*

Shadowfin soldierfish, *Myripristis adusta*

Seeing the light

In comparison with the other sighted subjects of this book, sea turtles need to see both above and below the surface. Their eyes are optimised to see well underwater and they are somewhat myopic in air. The green and hawksbill are the species most frequently encountered in the Coral Triangle. Both start life as opportunistic omnivores but upon maturity, the green turns vegan and is the only herbivore among sea turtles. The hawksbill remains omnivorous and will enjoy the epicurean delights of jellyfish, sea urchins, ascidians, molluscs and crustaceans. However, its preferred fare is specific species of sponges. Such delicacies are typically found at the base of dense stands of *Acropora* corals. The sequence shown here illustrates how irresistible these morsels are to the hawksbill and the energy that the turtles will invest to reach them. Compared to the gentle plucking of seagrass by the green turtle, the hawksbill's table manners involve plunging headfirst into a truly abrasive environment. This would imply a significant risk of damage to their eyes. The top left photo opposite provides an answer as the eye is seen to be virtually closed. It underlines the benefits of having three eyelids. The turtle may adequately see through the translucent nictitating inner third eyelid via the slit between the outer pair thus providing extra protection. A comparison of the heads of the two species strongly indicates the more abrasive lifestyle of the hawksbill.

Hawksbill turtle,
Eretmochelys imbricata

Hawksbill turtle, *Eretmochelys imbricata*

Green sea turtle, *Chelonia mydas*

CHAPTER 6

DECAPODS

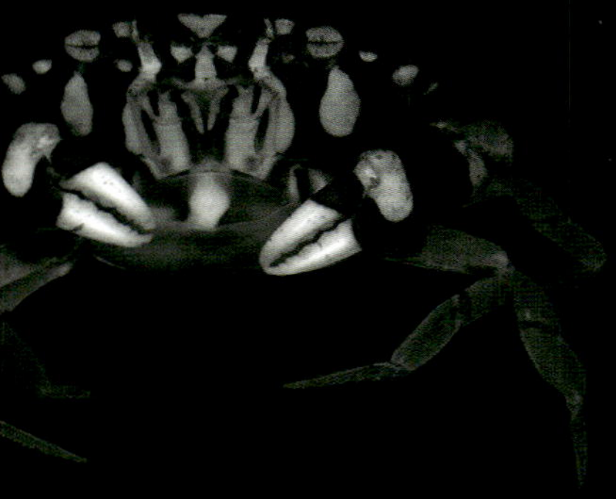

Zebra crabs, *Zebrida adamsii* on fire urchin, *Asthenosoma varium*

Emperor shrimp, *Zenopontonia rex* on sea cucumber

Easy riders

Decapods, particularly shrimps and crabs, enjoy a fascinating range of symbiotic relationships with other reef invertebrates. The zebra crabs on the chapter page find sanctuary among the venomous spines of the fire sea urchin as do the Coleman shrimps pictured on the following pages. The urchins also follow a nomadic existence as they wander over the substrate grazing on algae and thus can introduce their associates to new sources of food and encounters with potential mates. Emperor shrimps similarly enjoy the benefits of a mobile home by hitching rides clinging to the sides of holothurians, known as sea cucumbers. From such vantage points, they can be seen almost 'hanging ten' holding on with their back legs while reaching out with their pincers. From this mobile platform, they can thus pluck delicacies from the passing smorgasbord on the substrate below. Apart from dining on this external fare, these commensal shrimps also provide a bespoke service to the cucumber by consuming its parasites. As is evident in the examples shown, these tiny shrimp, measuring less than 2 cm, nicely blend into the patterned surface of the cucumber. Yet, in combination with plainer species of holothurians the shrimp actually stand out, so a lack of camouflage does not seem critical. They also turn hitchhiker to share commensal associations with various species of nudibranchs and thus derive similar mutual benefits.

Peacock-tail anemone shrimp, *Ancylocaris brevicarpalis*

Emperor shrimp, *Zenopontonia rex* on sea cucumber

Harlequin shrimp,
Hymenocera picta

Coleman shrimp,
Periclimenes colemani
on Fire urchin,
Asthenosoma varium

Coleman shrimp, *Periclimenes colemani* on Fire urchin, *Asthenosoma varium*

Working in pairs

Some renowned relationships of shrimps with other invertebrates are shown here and on the preceding two pages. Prominent among these are the Coleman and harlequin shrimps. They typically operate as a male and female team. Harlequin shrimp provide a macabre example of 'beauty and the beast'. Making no attempt to hide, a pair may be encountered boldly waving their seemingly delicate array of pastel-coloured appendages. Moving in a curious swaying motion, it is as if they are deliberately drawing attention to their lair. In fact it is an invitation to an invertebrate crime scene. For the uninitiated, it can be quite a shock to find themselves guests at a gruesome banquet. Harlequins slowly eat sea stars alive. Just one leg can last a fortnight. These diminutive shrimps are surprisingly adept at constraining their huge captives by simply turning them upside down. In a transparent comparison, the pair of shrimps shown opposite, display a clear delicacy in disguise. It is difficult to discern a pair as they completely merge into the surface of the mushroom coral. For added variety, an engaging example of shrimp and fish teamwork is included below. The ever vigilant goby has sharp eyesight enabling the myopic shrimp to continually toil away excavating the burrow of their shared accommodation. They will both rapidly retreat within if predators approach or, indeed, a hopeful underwater photographer. They are challenging subjects to portray.

Shrimp goby, *Amblyeleotris periophthalma*, with associate shrimp

Pair of cryptic shrimps, *Hamopontonia fungicola* on mushroom coral

Hermit crab variations

Hermit style

Hermit crabs are engaging characters often encountered scrambling over the reef at night when they emerge to feed. Daylight ventures are risky as predators such as parrotfish can make short work of a hermit despite its protective shell. They are not true crabs, whose abdomens are encased in their own hard exoskeleton, but are more related to squat lobsters. Hermits have soft and vulnerable abdomens — thus the vital need to acquire an abandoned shell. This provides portable protection into which they can fully retreat when threatened. Some species have been observed to show a creative community spirit when, at shell changeover time, they gather together in an ordered line from smallest to largest. They then perform a synchronised operation with each crab moving to a slightly larger shell. Of course it is critical that all shells are timely vacated. The individual opposite does not appear to share such niceties as the tightly closed operculum indicates that the original gastropod owner is still in residence. That direct and defiant glare betrays a growing frustration, which looks soon to be aggravated since its supplementary protection of sea anemones also needs to be accommodated. The individual on the left seems to have an improbably small shell but perhaps its ghostly glaze of bryozoans hold a special attraction. The example below pays testament to the extended service of a shell and a long heritage of previous owners. An impressive example of nature's recycling.

Hermit crab
with three tube
anemones
attached to shell

Corallimorph decorator crab, *Cyclocoeloma tuberculata*

Decorated deception

Decapod simply means 'ten footed'. Such anatomy is clearly evident in the crab opposite, despite its liberal decoration of stinging hydroids. This living array of potent protection is also a transferable asset. Remarkably, a moulting crab will painstakingly transplant its collection of hydroids to the new carapace. This is a delicate and precarious operation. It has to be undertaken during the period of heightened vulnerability while the crab's new exoskeleton hardens. The crab on the left has also adopted a singular approach to decoration, specialising in corallimorpharians, which most notably disguise the crab's outline. A different obscurity is achieved by the crab below which has a far more ad hoc approach to embellishment. Its eccentric arrangement of hydroids and sponges effectively combine to conceal the crab. Although its presence may be fairly evident in the photo, the angle was carefully chosen to best reveal its presence. From other perspectives it blended in completely and its disguise was only betrayed by a slight movement leading me to notice an eye. Decorator crabs have a covering of fibrous material armed with tiny hooks known as setae. These enable them to attach their adornments like Velcro. Further examples of creative flair in decapod decoration are provided on the following pages but, as noted in the next section, 'A tangled lineage', crabs seem to be the sole practitioners of this impressive art.

Cryptic decorator crab

Decorator crab,
Achaeus sp.

Teddy bear crab, *Polydectus cupulifer* carrying pair of anemones, *Phellia* sp.

Decorator crab, *Xenocarcinus tuberculatus*

Orangutan crab, *Achaeus japonicus*, on hard coral, *Fimbriaphyllia ancora*

Candy crab, *Hoplophrys oatesii*, on soft coral, *Dendronephthya* sp.

Hairy squat lobster, *Lauriea siagiani*

A tangled lineage

The impressive range of protective decoration used by crabs, formed from both the quick and the dead, is illustrated on previous pages. Yet, given that they have evolved in similar niches, it seems curious that there is no comparable proliferation among the thousands of species of shrimps and lobsters. The ad hoc collection of so-called lobsters presented here and overleaf marks further divergence. None are true lobsters. Despite their common names, they are not even near relatives. The squat lobsters are closer to hermit crabs — which themselves are not true crabs. Every time one lifts a metaphorical stone, another lineage seems to emerge. Wither the complex pathways of evolution. These 'pseudo' lobsters, display a wide spectrum of body size and form that differentiates them from true lobsters. In the spiny and slipper lobsters this is evident from the lack of claws on their front four pairs of (walking) legs. The former appear positively extrovert as they display their incredibly long and thick antennae. It is a testament to their spatial skills that they can successfully negotiate the complex topography of the reef with such extensive appendages. Squat lobsters have long, fine antennae and front legs, armed with substantial claws, that are longer than their bodies. The painted spiny and the sculptured slipper lobsters reach up to 40 cm and 20 cm in body size respectively, while the diminutive squat lobsters typically attain just 2 cm in length.

Sculptured slipper lobster, *Parribacus antarcticus*

Painted spiny lobster, *Panulirus versicolor*, below demosponge, *Petrosia* sp.

Crinoid squat lobster, *Allogalathea elegans*

Squat lobsters, *Galathea* sp. in sea pen, *Pennatula phosphorea*

Squat lobster, *Galathea* sp.

CHAPTER 7

GASTROPODS

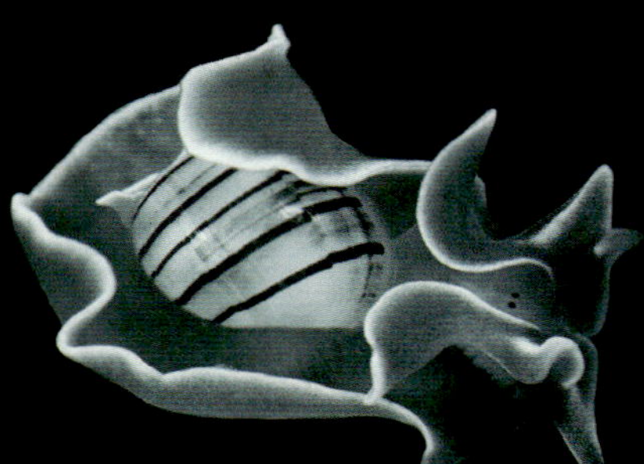

Rosy spindle cowrie,
Phenacovolva rosea

Ghost nudibranch,
Melibe cf. *engeli*

Fancy footwork

Sea slug is a rather prosaic term to describe members of such a major family within the phylum of Mollusca — especially as it includes many of the most exotic creatures to be found on the reef. They had tended to be overlooked by marine biologists but shot to prominence with the advent of scuba diving which enabled close observation in their natural environment. They are more fondly known as nudibranchs which means 'naked gills.' These 'branchs', which are located behind the heart, are exposed when their rudimentary form of a shell is lost at the end of the larval stage. While their delicate feathery gills can be withdrawn, the nudibranch relies, for its main protection, on its bright aposematic colouring to warn of its foul taste and toxins. Gastropod, literally translates as 'stomach foot.' This is an apt description for an animal whose stomach and foot are combined to form a common body part — a creative variation on putting a foot in the mouth. As highlighted in the examples here and on the following pages, evolution has also proved creative in the rich variation of texture, form and colour. Apart from their spectacular appearance, these fascinating animals have a surprising range of senses including sight, taste, smell and touch — if only in a fairly primitive form. Their short life cycles are compensated by both sexes being hermaphrodites, though they still need to mate since they cannot self fertilise (see example on page 7).

Nudibranch,
Arminidae sp.

Nudibranch,
Glossodoris atromarginata

Nudibranch, *Goniobranchus kuniei*

Nudibranch,
Ardeadoris averni

Nudibranch,
Janolus sp.

Blue dragon nudibranch, Aeolidae sp. laying eggs

Nudibranch,
Nembrotha yonowae,
consuming ascidian

A riot of rhinophores

Apart from their lurid colours and patterns, nudibranchs display a frilly protrusion of gills and rhinophores which decorate respective ends. At the front, the rhinophores, from the Greek for 'nose carrier,' protrude like tiny exotic horns. Visually, they are rather reminiscent of sporting a pair of miniature sea pens as head gear. There is a slight irony in this analogy since the Arminidae nudibranch, featured in 'Fancy footwork,' dines on sea pens. Others specialise in different fare. As may be deduced from the sense of intention, the *Nembrotha* pictured opposite feeds on ascidians. The rhinophores detect chemicals in the water and thus provide the nudibranch with the equivalent of a sense of smell. The reason for the rearing posture, shown below, is not clear but such elevation may assist in enhancing detection of chemicals in the current above the boundary layer. As these delicate structures may present a tempting snack to predators, they can be withdrawn like the gills. The slow moving nudibranch is not noted for its fast reflexes, so it would hardly react quickly enough to avoid a passing fish. For this, the aposematic deterrent is the main defence. Most nudibranchs have, at best, a very limited visual sense. They are only able to perceive variations in light intensity through tiny eye patches located near the rhinophores. Clearly then, their colourful livery plays no role in mating displays as they are sadly deprived of seeing what a graphic piece of artwork they are.

Nudibranch,
Nembrotha cristata

Nudibranch,
Nembrotha sp.

Nudibranch,
Nembrotha chamberlaini

Nudibranch,
Nembrotha kubaryana

Nudibranch, *Costasiella kuroshimae*

Exotic sheep

There are over three thousand recorded species of nudibranchs and the majority live in the warm, shallow waters of the Coral Triangle. They range in size from a few millimetres up to the 30 cm attained by the Spanish dancer, which weighs in at an impressive 1.5 kg. The subjects here are right at the other end of the scale with the *Costasiella* species measuring only 7 mm. They make their homes on the algae, *Avrainvillea*, upon which they graze, making for a conveniently self-contained lifestyle. Being herbivorous, they have the most unromantic common name of sap-sucking slug. As a counter to such a mundane moniker, these tiny exotica are fondly referred to as 'Shaun the Sheep,' made endearingly famous by the animated character. It is easy to appreciate why, as they do convey that baleful look from a decidedly sheep-like face. In practice such detail is only revealed by macro photography which, at this scale, is a challenge in itself. Their diminutive size and colouration makes them hard to spot and study with the naked eye. In a further dimension to their charm, they are solar powered, supplementing their diet by photosynthesis. This is achieved by absorbing the algae's chloroplasts within the tissues of their cerata — the green horn-like organs covering their backs. There are at least three different species represented among the six individuals shown here. One, opposite, is seen busily extending its household, laying down its tiny eggs in an intricate spiral.

Nudibranch, *Costasiella usagi*

Nudibranch, *Costasiella* sp. on *Avrainvillea* algae

Opisthobranch,
Micromelo undatus

Ghosts in the shell

These delightfully diaphanous species are commonly known as bubble snails. This relates to the thin bubble-like shell that they carry as an evolutionary souvenir which is far too small to accommodate its owner. Consequently, the protective operculum, which closes the entrance to a shell, has also been lost during the journey through deep time. However, this apparent fragility is misleading since, despite appearances, bubble snails are robust and energetic carnivores. They are not encountered on the coral reef proper, but inhabit the adjacent sandy or rubble areas. As they typically emerge to hunt after dusk, it is a special treat to find them active during the day. Confounding any preconceptions pertaining to mere slugs, they move with surprisingly alacrity. Like phantoms from another world, they seem to glide ethereally above the substrate as ghostly apparitions. Polychaete worms form their main diet which, ironically, also provide a source of defence. The snails incorporate the worms' toxins for their own protection. They will also consume other molluscs including nudibranchs. Bubble snails may be observed tracking the scent trail left in the sand by their slower moving prey. With a rippling flutter of their mantle, they will promptly disappear into the sand if their quarry attempts such an escape path. Examples of gastropods that enjoy fully protective shells are provided on the following four pages.

Opisthobranch
unidentified

Opisthobranch, *Hydatina physis*

Tiger cowrie,
Cypraea tigris

Pink-mouth egg cowrie,
Ovula costellata

Beautifully-banded moon snail, *Natica euzona*

Tapestry turban,
Turbo petholatus

CHAPTER 8

BIVALVES

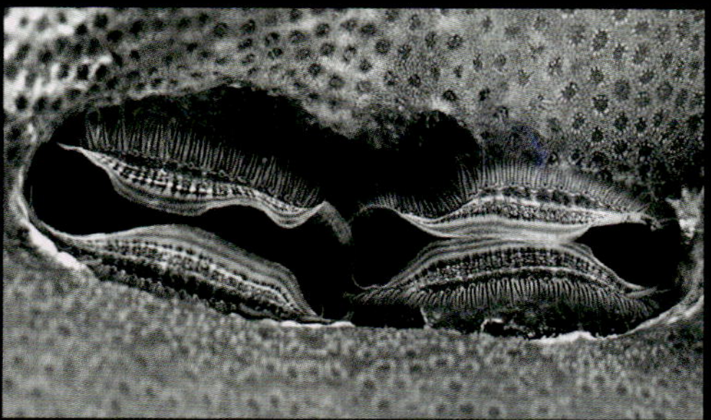

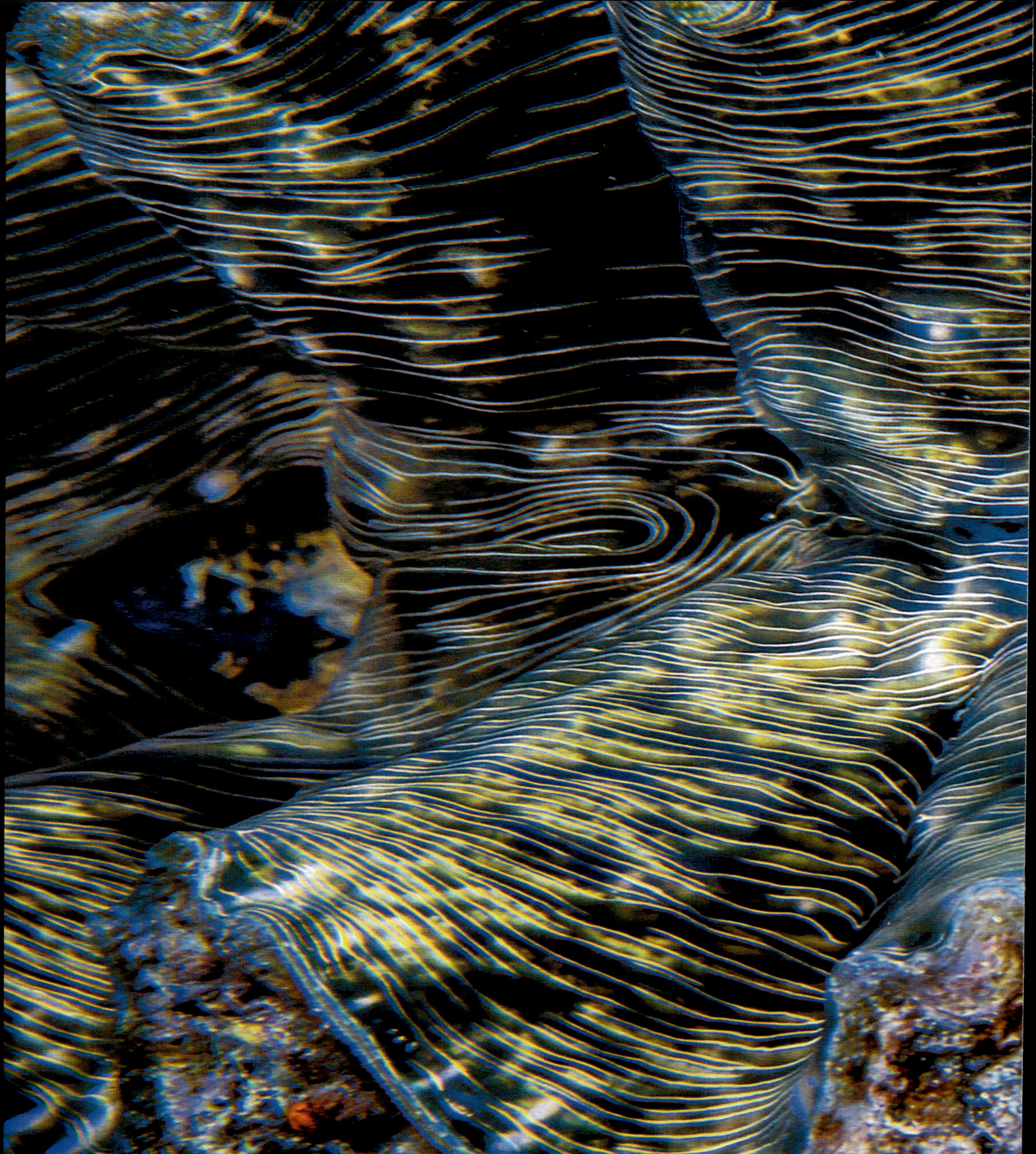

Giant clam,
Hippopus hippopus

Liquid art

Giant clams are the world's largest living bivalve molluscs. One *Tridacna gigas* specimen had a recorded length of 1.37 m and an estimated weight of around 250 kg. But it is not these impressive measurements that are the main attraction here. Rather, the chapter page and the six following illustrate the vibrant palette of iridescent colours and patterns that the various species exhibit in the mantle. Nestling among the corals, which provide a colourful setting in themselves, tridacnid clams heighten the sensations further. They are often encountered in relatively shallow water on the reef crest and may thus provide an enthralling location for a decompression stop. Sadly, such proximity to the surface makes them susceptible to widespread collection. They are keenly sought as food throughout the Indo-Pacific and are consequently on the International Union for Conservation of Nature Red List of endangered species. One of the most exquisite, *Hippopus hippopus*, is also one of the rarest. The clams themselves enjoy two sources of nourishment. While they filter feed nutrients such as plankton, their main source of energy is from photosynthesis. Like corals this derives from zooxanthellae embedded in the tissue of the mantle. This is a unique distinction among bivalves and why they favour shallow locations exposed to bright sunlight. Consequently, those intoxicating kaleidoscopes of colours are the result of symbiosis — the teamwork between microscopic algae and giant clams.

Giant clam,
Tridacna squamosa

Pattern variation in mantles of *Tridacna squamosa*

T. maxima

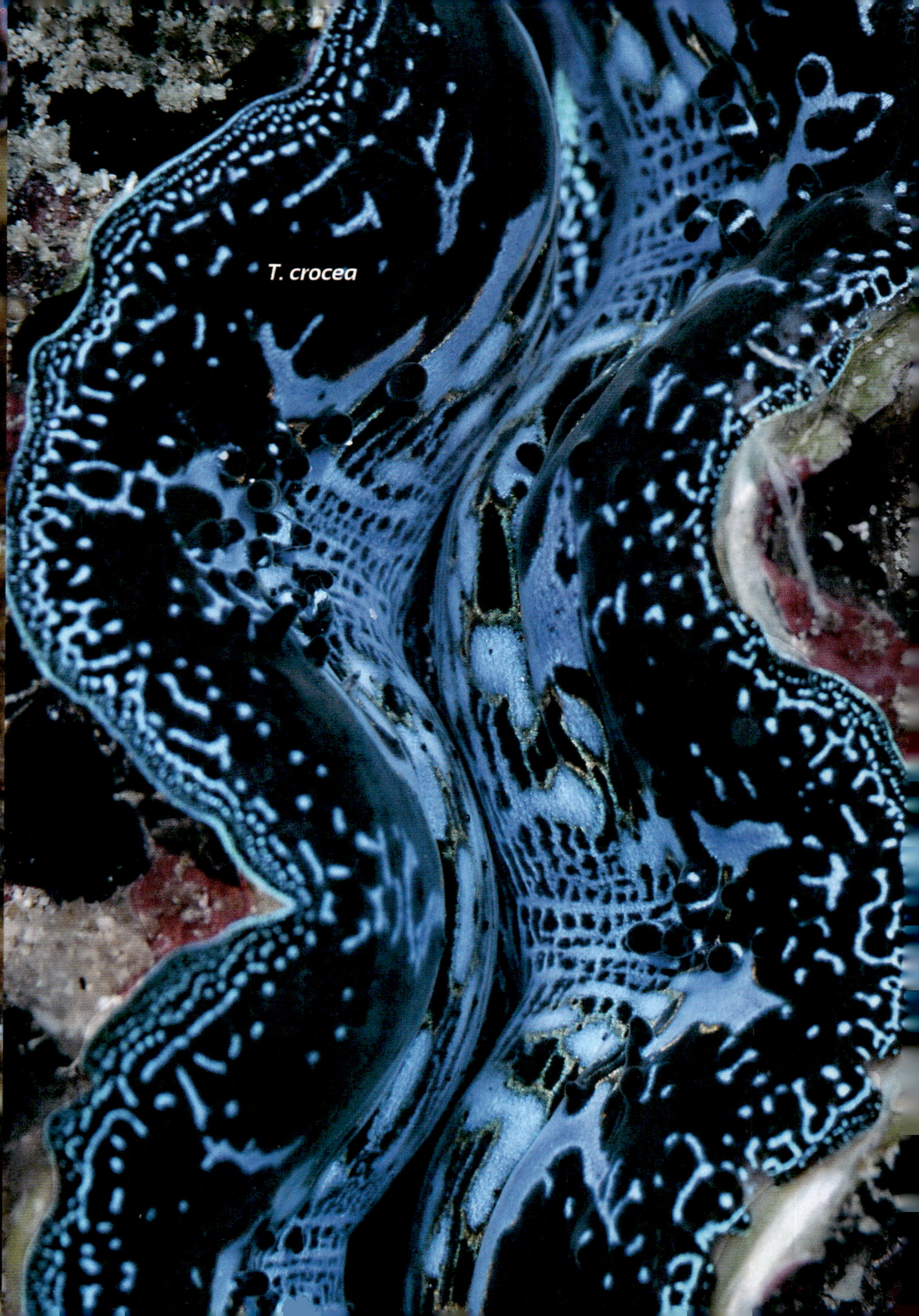
T. crocea

H. hippopus

T. maxima

Giant clam, *T. gigas*

An open and shut case

Although on a smaller scale, spiny oysters, like tridacnid clams, are also consummate practitioners of mantle display. The examples here and on the following pages showcase the range of these colourful creations in pallial art. However, the reasons for the richness of these exquisite tapestries are less clear. Unlike the tridacnid clams, the oysters do not enjoy a symbiotic relationship with a single-celled algae, but fully rely on filter feeding and are typically found in deeper and more shady retreats in the reef. They are deceptively complex creatures, boasting a range of organs including a heart, kidneys, and an extensive nervous system — but no brain. Like scallops, as discussed in *Chapter 5* 'In the blink of an eye,' many have surprisingly complex eyes. These are evident lining the mantle margin of the orange *S. varians* specimen overleaf. They are very sensitive to approach which, of course, makes for an engaging challenge to get a close macro shot. You only get one chance as they rapidly snap shut at any perceived threat. As can also be seen, their external surfaces attract a veritable cornucopia of encrusting organisms such as sponges and ascidians.

Thorny oyster, *Spondylus varians*, closed

S. varians
showing mantle

S. *varians*, open

Bivalve unidentified

S. *varians*, open

S. varians, open

Pearl oysters, *Pinctada* sp.

Oyster adornments of a different kind

The three species of oyster shown here were encountered high on the reef attached to soft corals. Here they are well placed to access passing nutrients. Their gills have rows of fine hair-like cilia to filter their planktonic food. Such exposed locations also enhance broadcast spawning events. Hopefully, for the pearl oysters on the left, such missions have been successfully concluded, dispatching their legacy of larvae to settle and colonise other reefs. Though each is now but an empty pair of shells, they remain as attractive decorations, providing a base for further adornment by encrusting invertebrates.

The pair of orange scallops below are well established on the sea fan. Here they compete for plankton not only with the coral but also the mobile feather stars, one of which displays the enterprising tactic of sitting directly over the scallop's opening. The honeycomb oyster opposite is still clearly with the living, both internally and externally, and is adorned with a riot of organisms all over its available surfaces. The aperture is deeply scalloped in a zig-zag pattern. To us the appearance of such a jagged looking mouth may appear threatening but its function is to enhance the flow of water and thus nutrition. The iridescent scallop and electric fire-clam shown overleaf complete the line-up of bivalve exotica.

Orange scallop, *Laevichlamys squamosa*

Honeycomb oyster, *Hyotissa hyotis*

Iridescent scallop,
Pedum spondyloideum

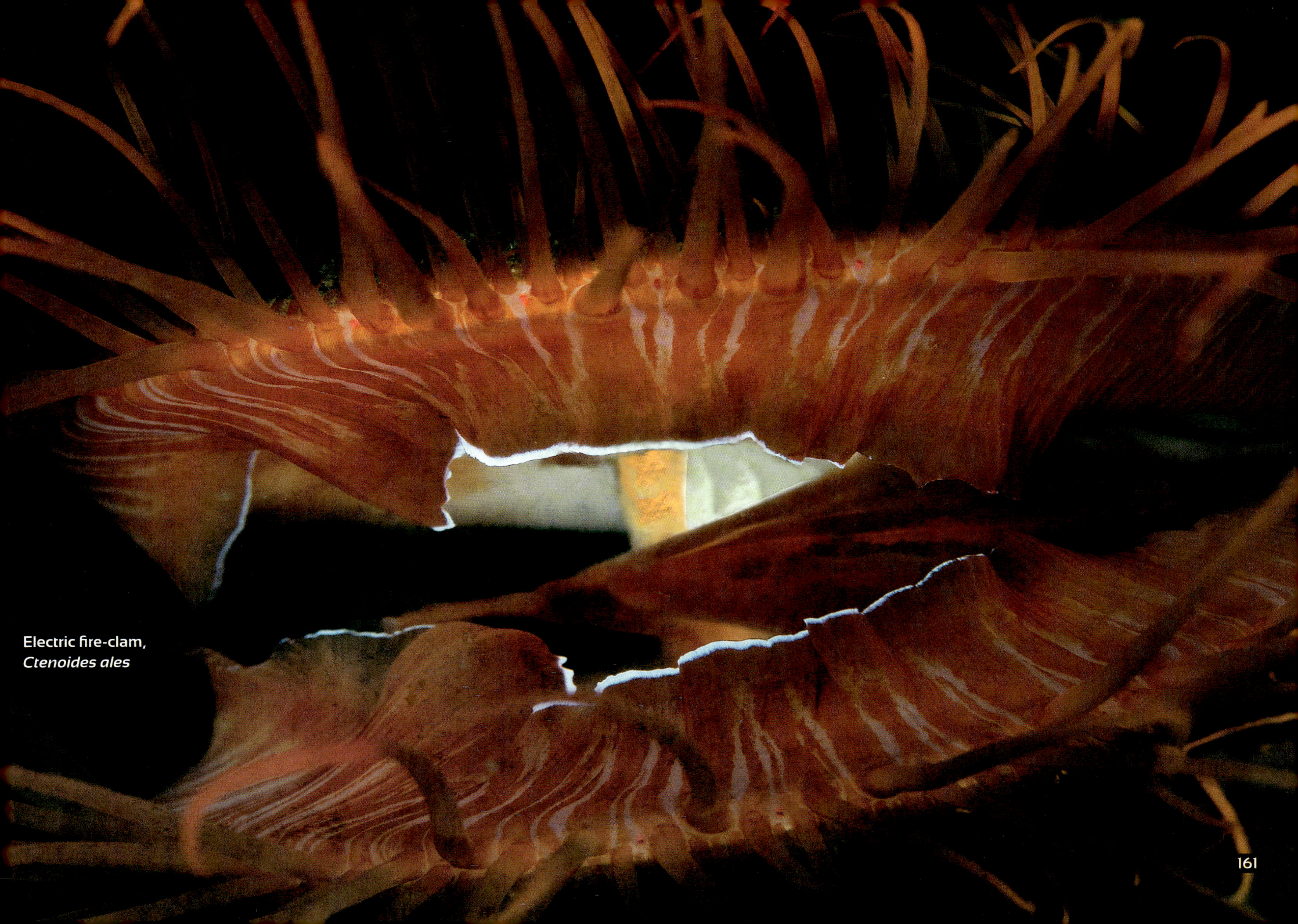

Electric fire-clam,
Ctenoides ales

CHAPTER 9

MARINE WORMS

Tube worm,
Sabellastarte sanctijosephi

Festive spirals

Compared to other Sabellida species, which emerge from their tubes like delicate flowers, the so-called Christmas tree worms bring a decidedly festive feel to the reef. They decorate the coral like brightly coloured miniature pine trees. In fact these exquisite creations are just the protruding gills as most of the worm is protected within a calcareous tube embedded in the coral. Embedment is a slow process but is achieved with minimal effort by the worm. It simply lays down the basic tube on the coral surface. The tube is then progressively entombed by the overgrowth of coral polyps. The gills, as shown, come in matching pairs, so one gets two trees for each worm. They are impressively multipurpose for, apart from respiration, the gills act as filters to trap food and, remarkably, enable some sense of vision by detecting changes in light through tiny eye spots called ocelli hidden amongst the fronds. The worms are super-sensitive to approach and will rapidly retract, slamming the entrance shut with the operculum. Examples of these trap doors, poised ready to close, may be seen below, as the orange edged red disc, and opposite. It is therefore quite a feat of familiarity for the blenny shown opposite to nestle among the spirals. The worms are long living, three decades or more has been reported. So in that regard, it is always Christmas somewhere on the reef. Other examples of sabellid reef art are shown on the following pages. These worms may look fragile but are deceptively robust.

Christmas tree worm, *Spirobranchus* sp.

Banda coral blenny,
Ecsenius bandanus

Magnificent tube worm, *Protula bispiralis* nestling among anemones

Tube worm,
Sabellastarte sp.

Spirobranchus sp.
detail

Sabellastarte sanctijosephi, detail

Polyclad flatworm,
Nymphozoon orsaki

Flat undulations

Over twenty thousand species of flatworm have been described and the vast majority are internal parasites such as flukes and tapeworms. Marine flatworms, however, do not parasitise but, although quite primitive, are free-living spirits. They have no internal body cavity and very few organs. Their high surface to volume ratio, enables them to breathe underwater by simple diffusion of gases. They extend their stomachs to excrete digestive juices allowing them to consume their food externally. The tiny Acoel flatworms can infest the surface of corals or even nudibranchs, as shown overleaf. However, the larger species hardly have a flat existence, but enjoy particularly colourful lifestyles. Being simultaneous hermaphrodites, they are both male and female. In the fiercely competitive world of the coral reef, this doubles the odds of successful reproduction. This evolutionary strategy does bring some interesting complications. Their eggs are fertilised and developed internally. An immediate conflict arises since any pair of flatworms of the same species can each deliver or receive sperm. Growing eggs is energy intensive, so each worm seeks to avoid such commitment and act as the male. To achieve this, some species have evolved the dramatic ability to use their penis to inject sperm directly through the body surface by 'hypodermic insemination'. In this battle of the sexes, known as penis fencing, the first to impale wins, and thus evades the demands of motherhood.

Polyclad flatworm,
Pseudoceros dimidiatus

Polyclad flatworm,
Phrikoceros baibaiye

Acoel flatworms, *Waminoa* sp., on nudibranch, *Halgerda batangas*

Acoel flatworms, *Waminoa* sp., on bubble coral, *Plerogyra sinuosa*

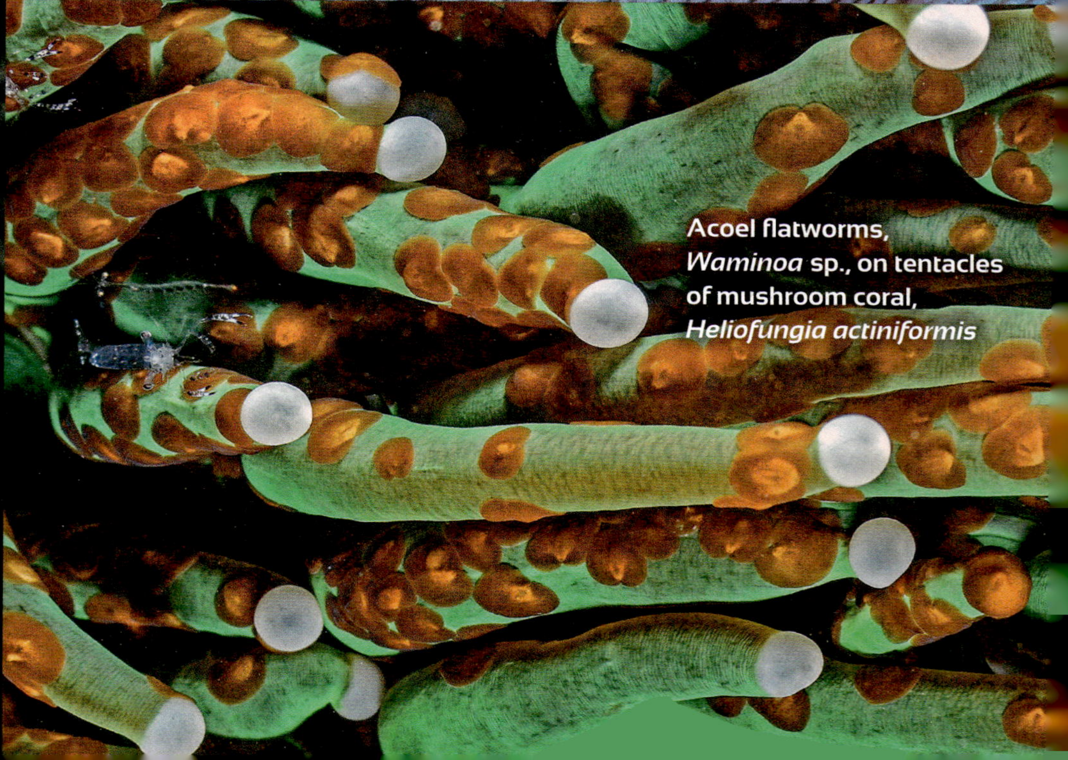

Acoel flatworms, *Waminoa* sp., on tentacles of mushroom coral, *Heliofungia actiniformis*

Persian carpet flatworm,
Pseudobiceros bedfordi
swimming

CHAPTER 10

SPONGES

Sea anemone,
Radianthus magnifica
in demosponge

Lionfish, *Pterois volitans* in demosponge

A not so simple life

Sponges are one of the most ancient forms of multicellular life and, before the 19th century, were not even recognised as animals. They have the most basic of body plans, devoid of tissues, organs, muscles and nervous system. Yet, having existed for over 600 million years, they are seriously successful. They number over ten thousand species, which populate the shallow seas in an abundant profusion of shapes and sizes. They have even been found thriving around deep sea hydrothermal vents. Their success relies on just three types of specialised cells which act in concert to pump, filter and digest nutrients. This option for a really simple life would seem an unlikely contender to provide a rich vein of study. But in fact sponge-related research is proving revelatory. Sponges are one of the most ancient forms of life to have their genomes sequenced. That trio of cells is helping to solve one of nature's great mysteries — the leap to multicellular life. Their genomes have provided a deep dive into the genetic processes that initiated teamwork between individual cells. Incredibly, a recent study has revealed sponges to have around 26,000 genes. This exceeds those of most other animals and is nearly double that of humans. That indicates a huge redundancy in evolutionary pathways. These radiate from the common ancestor from which all living animals — including sponges and humanity — evolved. This ancient shared predecessor incredibly contained the material to provide the genetic components for all complex life.

Apart from that prodigious genetic tool-kit of heredity, sponges may have other claims to fame. Recent studies have indicated that, although they have no nervous system, they have cells that may have made a vital contribution to the evolution of the nervous system. Sponges also hold the record for animal longevity with individuals estimated to live for eleven thousand years. Given that they have no immune system, this should render them especially vulnerable to cancers, but in fact they can survive huge exposure to radiation without damage and are now being studied in research for cancer treatment.

It was previously considered that the symbiosis between corals and their unicellular algae was the prime example of generating energy through photosynthesis in the marine world. This relationship dates back over 200 million years to the Triassic and provided the foundations for the immense diversity of modern coral reefs. However, current research is indicating the possibility of a much earlier progenitor of such significant teamwork. This relates to an analogous photosynthetic symbiosis between sponges and the cyanobacteria that live within them. Incredibly, this partnership, which is over 500 million years old, may have played a key role in generating the Cambrian explosion by converting the deep oceans from an anaerobic wilderness to an oxygen-rich environment. This continued to sustain the sudden flowering and radiation of the vast range of animal life we know of today including, of course, the subsequent evolution of coral reefs. The aggressive acquisition of such real estate is clearly demonstrated by the reef fish and the chapter page features an anemone adopting an encroaching sponge as a nesting site. Competition abounds.

Moorish idol,
Zanclus cornutus
Barrel sponge,
Xestospongia sp.
Emperor angelfish,
Pomacanthus imperator

Moray eel,
Gymnothorax fimbriatus
among encrusting sponges

Sponge crab, *Dromia* sp.
with sponge, *Clathria reinwardti*

Space wars

In the incredibly competitive world of the tropical coral reef, the winning and holding of space is a dominant factor. This struggle is vividly demonstrated overleaf where different sponges compete together with corals and fish. A vast range of tactics have evolved and the cogency of the 'Red Queen' hypothesis is ever evident: 'It takes all the running you can do to keep in the same place'. Among the impressive list of contestants, sponges are apex predators of space. After all, they have had plenty of practice having been honing their prowess since early Phanerozoic times. Boring sponges use acids to penetrate stony corals. For example, in undermining the base of the fast growing stands of *Acropora* corals, they play their part in the natural cycle of collapse and regrowth of these common reef structures. Sponges have also perfected the art of overgrowing and encrustation. As the examples in this chapter show, other species extend like tentacles to embrace and overwhelm competitors. Studies have revealed that sponges can progressively rearrange their internal anatomy and this facility to remodel may enable them to physically change shape, thus giving them a further edge in the battle for living space. It has also been found that collective mobilisation by tiny spicules on its outer surface can enable an entire sponge to move, if only at a modest rate. The swiftest can achieve around 4 mm a day which is no mean feat for a supposedly sessile animal.

Blue sponge, *Gelloides* sp. overgrowing sea fan, *Melithaea* sp.

Red tube sponge, *Axinyssa* sp. overgrowing hard coral, *Diploastrea heliopora*

Invasive orange
demosponge

Blue (*Gelloides* sp.) and orange (*Acarnus erithacus*) demosponges invading base of hard coral, *Acropora* sp.

Elephant ear sponge,
Ianthella basta

CHAPTER 11

ECHINODERMS

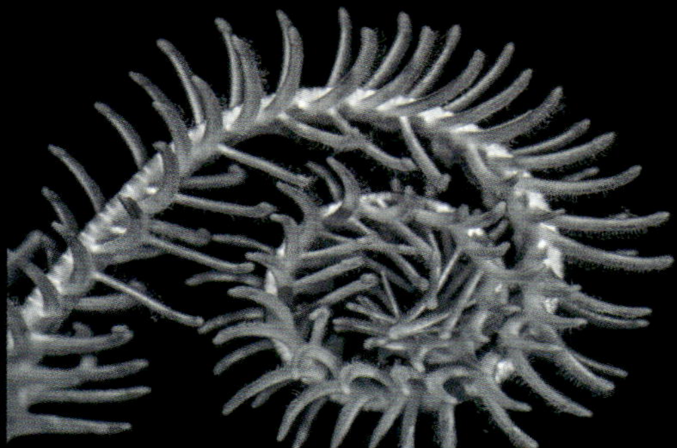

Feather stars (or crinoids), *Clarkcomanthus* sp.

Crinoid shrimps, *Laomenes* sp.

A welcome with open arms

Crinoids or feather stars belong to the phylum Echinodermata. Their ancestors first appear in the fossil record of the Cambrian period around 530 million years ago, making them some of the most ancient of marine animals. They flourished up to the Mesozoic era but were almost completely wiped out during the largest mass extinction event recorded some 250 million years ago. Just a few species are known to have slipped through yet, while not regaining their former glory, they have rebounded to boast some 550 species now extant today. That is quite a triumph for such inherently delicate-looking creatures. Their arms do readily break off but this apparent fragility is compensated by their ability to grow replacements. Indeed, a new arm may be seen growing on the *Clarkcomanthus* opposite and specimens have been recorded sporting an impressive array of over 150 arms. This seems at odds with the five-fold symmetry that they share with all other echinoderms. In fact, as shown overleaf, the arms do radiate in multiples of five from a central disc. They then fork extensively to create the appearance of abundant numbers of separate arms.

As featured in this chapter, the extended period of evolution enjoyed by crinoids, has enabled a colourful range of symbiotic relationships to develop, particularly with decapods.

Crinoid shrimp with squat lobster, *Allogalathea elegans*

Criniod,
Clarkcomanthus sp.

Pair of snapping shrimp, *Synalpheus* sp. with blue *Myzostoma* worm in crinoid

Colourful guests

Crinoids provide a most attractive haven for a wide variety of tiny creatures, including shrimps, squat lobsters and worms. Commensal shrimps, in particular, have evolved patterns and colours that closely match their crinoid host, making them nicely camouflaged to wander about along those outstretched arms. These have channels known as ambulacral grooves along which the planktonic nutrients are conducted downwards towards the mouth. The shrimp provide cleaning services to their host by consuming any debris that gets snagged in those delicate conveyor belts. Thus both animals benefit. Being opportunists, the shrimp also help themselves to the stream of food particles snared by their host. The filter feeding crinoid mouth shares the same location as their anus, on the upper surface of the body. This anatomical arrangement differs from other echinoids where, as they are grazers, by necessity the mouth is located on the underside. As shown, their undersides have claw-like appendages which enable them to secure a foothold on any convenient substrate including such prime feeding locations at the tops of sea fans. The pair of commensal shrimps opposite provide a further example in the art of blending in, as do the Myzostomid worms. The latter are usually found around the crinoid's central disc and often, as shown below, display matching cryptic livery.

Colour variations in *Myzostoma* worms

Pair of crinoid shrimps, *Laomenes* sp. at centre of host

Unwinding arms of crinoids, *Clarkcomanthus* sp. at night

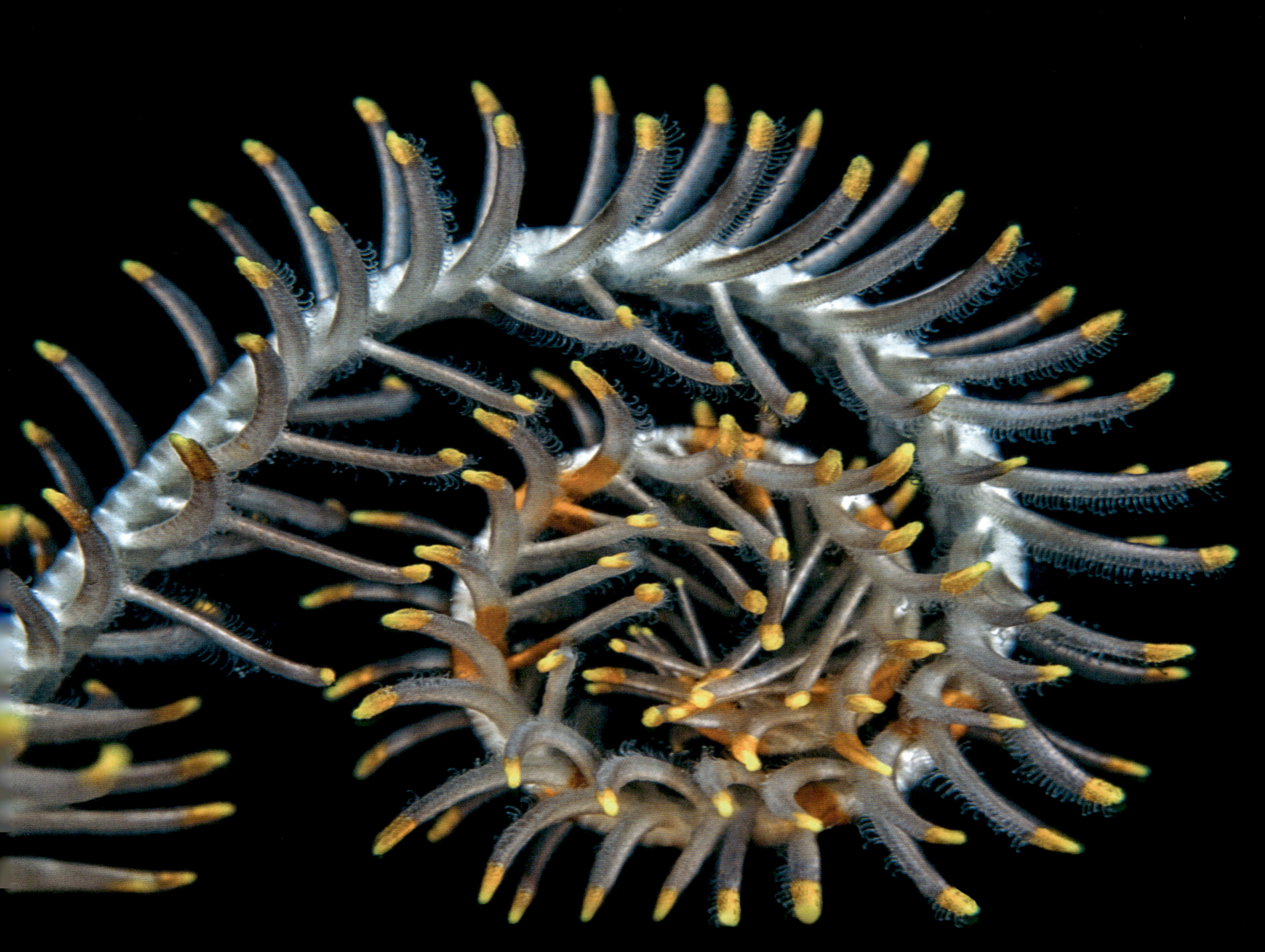

Clarkcomanthus sp.

Time to unwind

As highlighted by the dramatic splash of colour of the trio on the chapter page, crinoids, although predominantly nocturnal, may be active during the day. This somewhat risky opportunism is typically stimulated by periods of plankton abundance. This is readily evident from hordes of anthias and other planktivorous fish joining in the banquet high in the water column. Such bounty can thus provide enough distraction for the crinoid's erstwhile predators. All this feasting encourages crinoid symbionts to emerge and it is a real pleasure to be able to closely observe them during these times. Squat lobsters are also shown enjoying this liberty in *Chapter 6*. The symbionts' camouflage enables them to hide in plain sight while exposed way out on crinoid arms. Nevertheless, they remain a challenge to spot and are ever ready to disappear deep within the arms. While this may endow these symbionts with the confidence to forage so openly, it is dependent on the disposition of their hosts. If the arms are not extended then the symbionts must stay within the embrace of closed arms. The onset of darkness stimulates the crinoids to fully emerge and literally unwind. With their daytime predators now dormant within the confines of the reef, the crinoids join the nocturnal banquet. They slowly uncoil, converting their tightly closed balls of arms through decorative spirals until fully open. The phases of this mesmerising display are conveyed on the surrounding pages.

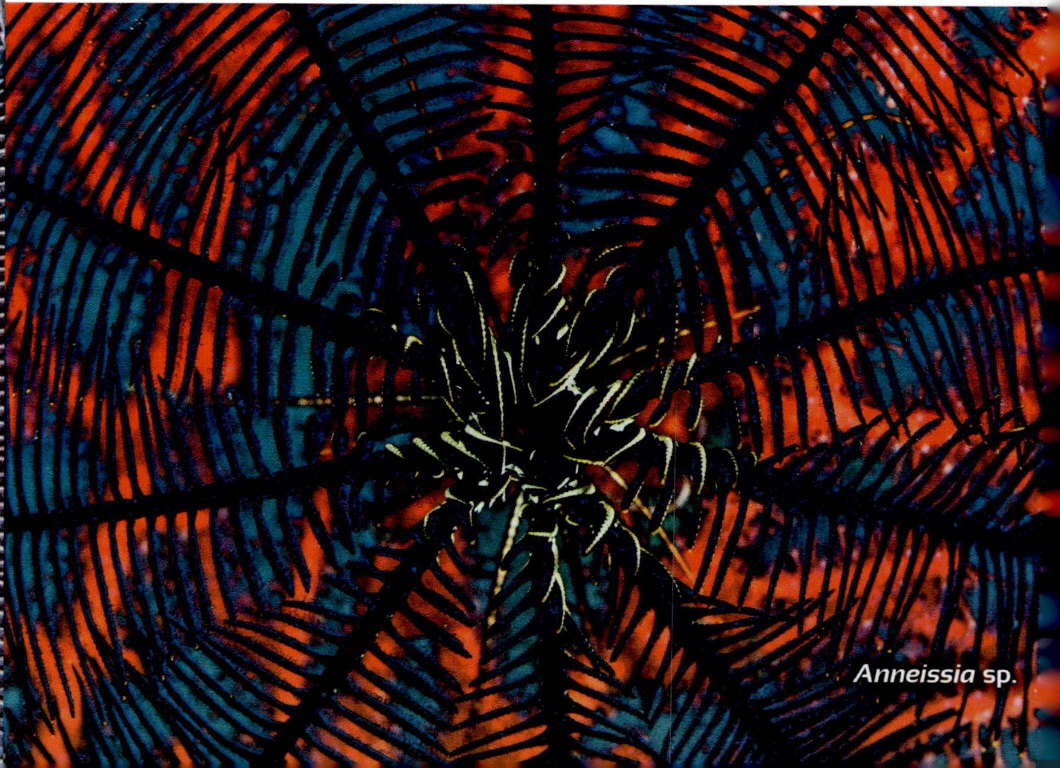

Anneissia sp.

Unwinding
arms of crinoid,
Clarkcomanthus sp.

Closed and open examples of crinoids, *Clarkcomanthus* sp.

Sea star mouth,
Fromia sp.

Warty sea star,
Echinaster callosus

Starry nights

Even more than crinoids, brittlestars are predominantly creatures of the night. While they share a marked propensity for arm loss, the aptly named brittlestar is an exemplar of creative fragility. Moreover, they do not generate veritable bushes of arms through progressive branching like crinoids, but are generally confined to the basic five. Even so, their arms do indeed easily break off. Such a lack of redundancy would hardly seem compatible with brittleness. Yet this apparent vulnerability has evolved to deliver defensive and reproductive benefits. If faced with imminent predation, sophisticated neural pathways enable the brittlestar to promptly detach an arm, thus distracting the predator while the amputee makes safe its escape. Perhaps even more bizarrely, like some fancy of scientific fiction, body fragmentation also allows brittlestars to reproduce asexually. This is achieved through a cloning process, known as fissiparity, in which the separated parts proceed to generate individually into complete stars. This complex durability is underlined by brittlestars having the most species of all echinoderms — well over two thousand. They inhabit diverse habitats from the equator to the polar regions, both shallow and deep. It is estimated that there are more than eight hundred species in the Coral Triangle.

Basket sea star, *Astroboa* sp., juvenile

Chocolate chip sea star, *Protoreaster nodosus* with juvenile brittlestars, *Ophiothrix* sp.

Brittlestar, *Ophiothrix* sp. on Fungiidae hard coral substrate

Brittlestar,
Ophiothrix sp.
on branching
Acropora sp.
hard coral

Urchin crab, *Dorippe frascone* carrying a radiant sea urchin, *Astropyga radiata*

Fiery havens

The Coral Triangle, harbouring the world's greatest biodiversity of marine life, was created in the volcanically active cradle of the 'Ring of Fire'. Evoking the resonant imagery of lava-bright fiery margins, it is tempting to view the sea urchins featured here as miniature analogues of tectonic plates. The radiant sea urchin, *Astropyga radiata*, is a common species in the Coral Triangle. Its genus name, deriving from the Greek, essentially means 'star-shaped anus'. More predictably it is known as the blue-spotted urchin and also the false-fire urchin. However, from direct experience, I can confirm that there is nothing false about their fiery nature. The close observer of benthic marine life may be unexpectedly vulnerable. It is the urchin's surprising mobility that can catch one out. Adopting a seemingly safe and spacious position on the sand may not remove the threat of an imminent impaling. The prone form of a diver seems to be quite an attraction and an urchin will promptly approach to nestle against the diver's side. Any inadvertent movement may then render the encounter unforgettable. The urchins can be even more disturbingly active at night. I had a very near miss with one almost landing on my head as it tumbled downwards in the dark from a coral overhang. I've worn a hood ever since, no matter how pleasantly warm the water may feel. Despite initial perceptions, the natural fauna of the reef, from fish to octopus and crabs, enjoy far more comfortable relationships with the urchins.

Juvenile Banggai cardinalfish, *Pterapogon kauderni*, sheltering in radiant sea urchin, *Astropyga radiata*

Veined octopus,
Amphioctopus marginatus

Sea urchin cardinalfish, *Siphamia tubifer*

Needle-spined sea urchin, *Echinostrephus aciculatus*

Zebra crab,
Zebrida adamsii, on
magnificent fire urchin,
Asthenosoma ijimai

CHAPTER 12

ASCIDIANS and other NVERTEBRATES

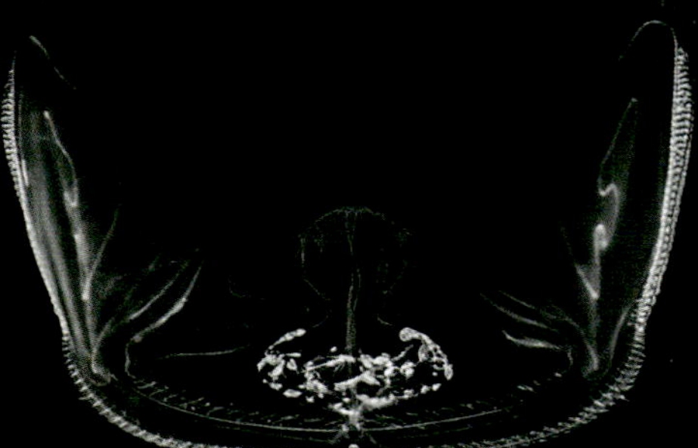

Ascidians,
Nephtheis fascicularis
and *Rhopalaea* sp.

Giant frogfish,
Antennarius commerson
on ascidians, *Clavelina* sp.

Fair and foul

Ascidians, also known as sea squirts or tunicates, embellish the reef in bright colours or glow with a subtle translucence. But, also being major contributors to marine fouling, they sadly suffer a Jekyll and Hyde reputation. The resulting negative economic impact presents a substantial threat to shellfish aquaculture farms. Their ability to overgrow is illustrated left where a marine anchorage is comprehensively festooned. This hazy décor is clearly welcomed by the resident giant frogfish as a deluxe ambush platform. In other examples of creative colonisation, the large ascidian below acts as substrate for a smaller species and the blenny opposite adopts another as a colourful lookout. Recent research has revealed that ascidians can fully regenerate from just a tiny segment using special stem cells in their blood. They spread extensively through budding but also develop eggs and sperm to reproduce sexually to enjoy an initial existence as free swimming tadpole-like larvae. All this reproductive repertoire is particularly impressive since they are biologically complex and are in fact our closest living relatives among invertebrates. Further examples of their integration in marine ecosystems are shown overleaf, ranging from fish egg depositories to nudibranch food banks and night time accommodation for pufferfish. Note that the nudibranch is directly consuming the insides of the ascidian thus avoiding the indigestible cellulose of the exterior tunic.

Ascidians,
Pycnoclavella sp.
covering
Herdmania sp.

Blenny, *Cirripectes* sp. on ascidian, *Polycarpa aurata*

Goby with eggs
on ascidian,
Rhopalaea sp.

Nudibranch,
Nembrotha chamberlaini,
consuming blue ascidian

Sharp-nose pufferfish,
Canthigaster rostrata,
asleep on ascidian,
Polycarpa aurata

Branching bryozoan, *Reteporella* cf. *graeffei*

Ciliated lacework

Bryozoans are an ancient life form which dates right back some 530 million years to the Cambrian explosion. They are a notably large phylum constituting some 6,500 living species. Colonial animals, bryozoans are predominantly marine and so abundant geologically that they have become important bio-markers in the fossil record. They exist in a range of delightful calcareous structures and are consequently also known as lace corals. Somewhat less poetically, their Greek name of Ectoprocta means 'outside anus.' The mouth opens into a U-shaped gut which loops to exit at the anus just outside the circle of tentacles. Perhaps surprisingly, their closest relatives are considered to be the brachiopods which, with their twin shells, are hardly a physical analogue. It may seem somewhat indulgent to devote this whole double-page spread to bryozoans, but they are often overlooked despite the delicate beauty of their various forms. Sadly, they are better known for negative reasons as they share a reputation with ascidians for fouling ships and marine infrastructure. They adhere to all manner of surfaces and a biological cameo of attachment is shown on the shell of a hermit crab on page 116. Bryozoans filter feed using a bell-like structure called a lophophore. This has a circular arrangement of tentacles around a central mouth. The tentacles, somewhat reminiscent of tiny cotton buds, are covered with hair-like cilia. These generate feeding currents to draw in suspended nutrients.

Bryozoan, *Triphyllozoon* sp.

Bryozoan,
Triphyllozoon sp.,
detail

Ascidian,
Rhopalaea sp.

Alien drifters

While the rich biodiversity of the Coral Triangle may readily overwhelm the senses, some bizarre transparent creatures may float by unnoticed. As illustrated, these range from ctenophores to heteropods and even pelagic ascidians. Ascidians are typically sessile creatures and, as noted, establish colonies that can spread extensively. The brief phase of the ascidian tadpole is totally focused on settling, affording no time to feed. Accordingly, they have no mouth but do possess a range of chordate components including a primitive spine, and rudimentary eyes and brain. After settlement, these redundant body parts are all absorbed in a fantastical transformation that matches the terrestrial example of the caterpillar to butterfly metamorphosis. But for the ascidian it is a case of losing rather than gaining mobility. The larva, now attached to the substrate, transmutes into a vase-like form for which it promptly develops intake and exit siphons to continue life as a filter feeder. Not all ascidians sacrifice mobility as there are pelagic species known as pyrosomes which may be seen rolling over the substrate like aquatic tumbleweed. Ctenophores are voracious predators and come in a range of shapes and body sizes. Like a tiny iridescent refugee from *Star Trek*, that opposite, with its gut full of plankton, has clearly enjoyed a successful nocturnal hunt. Overleaf, heteropods, curiously known sea elephants though actually molluscs, virtually defy description.

Colonial pelagic
ascidian,
Pyrosomatidae sp.

Comb jelly or
ctenophore,
Leucothea sp.

Sea elephant
or heteropod,
Pterotrachea sp.

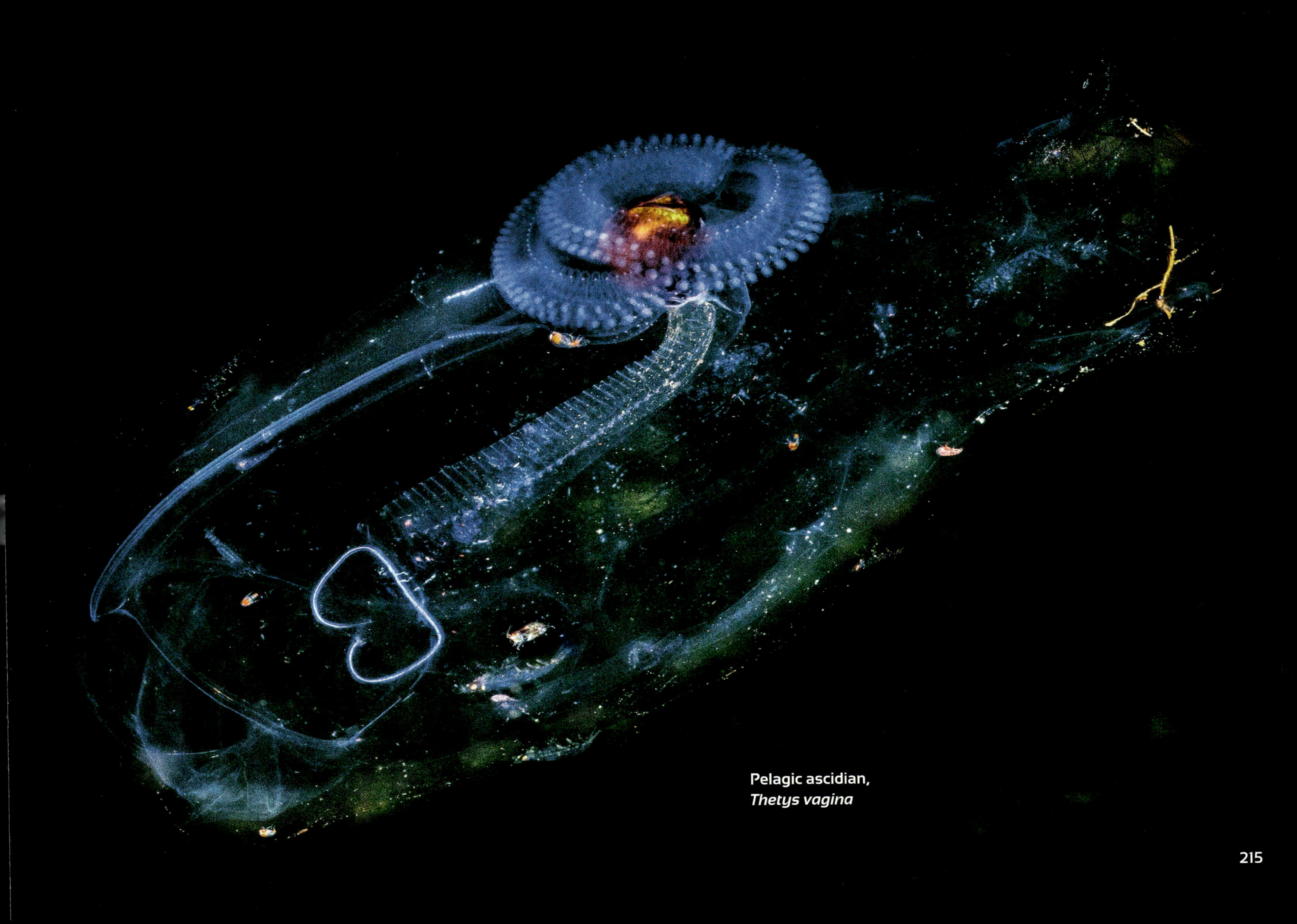

Pelagic ascidian,
Thetys vagina

Acknowledgements

We would like to thank the following for their expert advice on marine biology: Sammy De Grave, Charles Fransen, Harry ten Hove, Mei Lin Neo, Katja Peijnenburg, Noa Shenkar, and Nicole de Voogd. Special thanks are also due to Liz Powderham and Bastian Reijnen for their ongoing support.

Triton Mauwara, West Papua

Photo courtesy of Bruno Hopff, Amira

Equipment

Diving

Rebreathers: Poseidon Mark VI and SE7EN.
These are fully closed-circuit, fully automated systems.

Photographic

Cameras: Nikon DSLRs D7000 and D500.
Macro lenses: Nikon AF-S 105 mm f2.8G VR Micro, and Nikon 60 mm f2.8D Micro, supplemented with Subsee Underwater close-up lenses.
Wide angle lens: Tokina AT-X DX Fisheye 10–17 mm f3.5–4.5 zoom.
Underwater housings: Nauticam.
Strobes: Inon Z240 and S2000.

Further reading

Allen G.R. and Erdmann M.V. (2012) *Reef Fishes of the East Indies*, vol 1–3. Tropical Reef Research with Conservation International Indonesia.

Allen G.R. and Steene R. (2002) *Indo-Pacific Coral Reef Field Guide*. Tropical Reef Research, Singapore.

Behrens D.W. (2005) *Nudibranch Behaviour*. New World Publications Inc., USA.

Brandl S.J., et al. (2019) 'Demographic dynamics of the smallest marine vertebrates fuel coral reef ecosystem functioning.' *Science*, 364: 1189–1192.

Burke L., Reytar K., Spalding M. and Perry A. (2012) *Reefs at Risk Revisited in the Coral Triangle*. World Resources Institute.

Coleman N. (2008) *Nudibranchs Encyclopedia — Catalogue of Asia / Indo Pacific Sea Slugs*. Neville Coleman's Underwater Geographic Pty Ltd.

Colin P.L. and Arneson C. (1995) *Tropical Pacific Invertebrates*. Coral Reef Press, USA.

Debelius H. (2001) *Asia Pacific Reef Guide*. IKAN-Unterwasserarchiv, Frankfurt, Germany.

Fautin D.G. and Allen G.R. (1994) *Anemone Fishes and their Host Sea Anemones*. Tetra Press, Germany.

Gosliner T.M., Behrens D.W. and Williams G.C. (1996) *Coral Reef Animals of the Indo-Pacific*. Sea Challengers, USA.

Huang D., Goldberg E.E., Chou L.M. and Roy K. (2018) 'The Origin and Evolution of Coral Species Richness in a Marine Biodiversity Hotspot.' *Evolution*, 72: 288–302.

Humann P. and Deloach N. (2010) *Reef Creature Identification*. New World Publications Inc., USA.

Jones B. and Shimlock M. (2011) *Diving Indonesia's Bird's Head Seascape*. Saritaksu Editions, Bali with Conservation International Indonesia.

Kassem K. and Madeja E. (2014) *The Coral Triangle*. John Beaufoy Publishing Ltd., Oxford, UK.

Keenan G. (2016). 'Fish recorded singing dawn chorus on reefs just like birds.' *New Scientist*, www.newscientist.com/article/2106331

Kuiter R.H. and Debelius H. (2009) *World Atlas of Marine Fauna*. IKAN-Unterwasserarchiv, Frankfurt, Germany.

Le Page M. (2019) 'The tiniest fish are the most important for healthy coral reefs.' *New Scientist*, www.newscientist.com/article/2204346

Powderham A.J. and Van der Meij S.E.T. (2021) *At the Heart of the Coral Triangle — Celebrating Biodiversity*, CRC Press.

Ryanskiy A.S. (2018) *Coral Reef Crustaceans from Red Sea to Papua*. Reef ID books.

Sheppard C. (2021) *Coral Reefs — A Natural History*. Princetown University Press, USA.

Spotila J.R. (2005) *Sea Turtles — A Complete Guide to their Biology, Behavior, and Conservation*. Johns Hopkins University Press, USA.

Starling S. (2017) 'Fantastic beasts — cephalopod RNA recoding.' *Nature Reviews Genetics*, 18: 329.

Veron J.E.N., Devantier L.M., Turak E., Green A.L., Kininmonth S., Stafford-Smith M. and Peterson N. (2009) 'Delineating the Coral Triangle.' *Galaxea, Journal of Coral Reef Studies*, 11: 91–100.

Vagelli A.A. (2011) *The Banggai Cardinalfish — Natural History, Conservation and Culture of Pterapogon Kauderni*. Wiley-Blackwell, UK.

Yvaine Y. (2018). 'Coral likes to make its ocean home in places with noisy neighbours.' *New Scientist*, www.newscientist.com/article/2187928

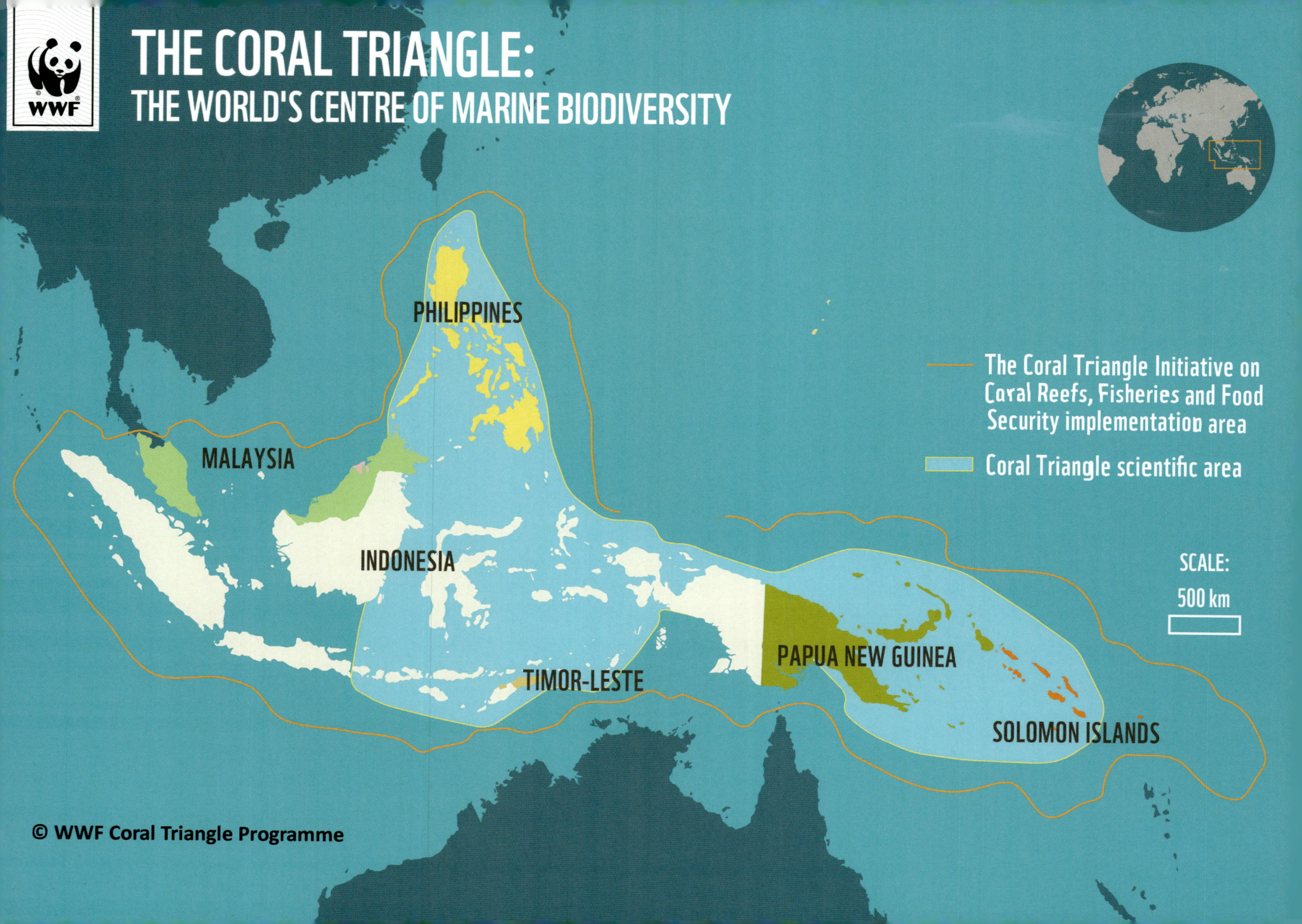

THE CORAL TRIANGLE:
THE WORLD'S CENTRE OF MARINE BIODIVERSITY
WWF
PHILIPPINES
MALAYSIA
INDONESIA
TIMOR-LESTE
PAPUA NEW GUINEA
SOLOMON ISLANDS
The Coral Triangle Initiative on Coral Reefs, Fisheries and Food Security implementation area
Coral Triangle scientific area
SCALE:
500 km

Index

D

E

F

G

H

I

J

L

M